T0177279

Textile Science and Clothing Technology

Series editor

Subramanian Senthilkannan Muthu, Bestseller, Kowloon, Hong Kong

More information about this series at http://www.springer.com/series/13111

Subramanian Senthilkannan Muthu
Editor

Detox Fashion

Sustainable Chemistry and Wet Processing

 Springer

Editor
Subramanian Senthilkannan Muthu
Bestseller
Kowloon
Hong Kong

ISSN 2197-9863   ISSN 2197-9871 (electronic)
Textile Science and Clothing Technology
ISBN 978-981-13-5254-6   ISBN 978-981-10-4876-0 (eBook)
DOI 10.1007/978-981-10-4876-0

© Springer Nature Singapore Pte Ltd. 2018
Softcover re-print of the Hardcover 1st edition 2018
This work is subject to copyright. All rights are reserved by the Publisher, whether the whole or part
of the material is concerned, specifically the rights of translation, reprinting, reuse of illustrations,
recitation, broadcasting, reproduction on microfilms or in any other physical way, and transmission
or information storage and retrieval, electronic adaptation, computer software, or by similar or dissimilar
methodology now known or hereafter developed.
The use of general descriptive names, registered names, trademarks, service marks, etc. in this
publication does not imply, even in the absence of a specific statement, that such names are exempt from
the relevant protective laws and regulations and therefore free for general use.
The publisher, the authors and the editors are safe to assume that the advice and information in this
book are believed to be true and accurate at the date of publication. Neither the publisher nor the
authors or the editors give a warranty, express or implied, with respect to the material contained herein or
for any errors or omissions that may have been made. The publisher remains neutral with regard to
jurisdictional claims in published maps and institutional affiliations.

Printed on acid-free paper

This Springer imprint is published by Springer Nature
The registered company is Springer Nature Singapore Pte Ltd.
The registered company address is: 152 Beach Road, #21-01/04 Gateway East, Singapore 189721, Singapore

This book is dedicated to:
The lotus feet of my beloved
Lord Pazhaniandavar
My beloved late Father
My beloved Mother
My beloved Wife Karpagam and
Daughters—Anu and Karthika
My beloved Brother
Last but not least
To everyone working in the global textile
supply chain to make it TOXIC FREE &
SUSTAINABLE

Contents

Sustainable Chemicals: A Model for Practical Substitution

Christina Jönsson, Stefan Posner and Sandra Roos

Abstract The textile industry sees currently a fast development of legal and voluntary restrictions of chemicals content in textile products. However, the on-going phase-out work focuses on evaluating the environmental and health aspects of chemicals. The technical performance in the end application for the chemical does not receive the same attention. In addition, many research projects committed to evaluating hazardous substances and their possible alternatives also neglects the technical performance. The technical performance is left to the companies to evaluate. This may lead to inefficiency in the substitution process and also have the consequence that companies never dare to take the step to practical substitution, at least not in a proactive way. This chapter presents a model for practical substitution, developed and evaluated in several case studies, whereof two in the textile field: water and soil repellent textile coating materials and flame retarded textiles. From the general lessons learnt, an improved substitution methodology with widespread applicability has been defined.

Keywords Chemicals · Toxicity · Practical substitution · Perfluorinated substances · Flame retardants · Functional properties

1 Introduction

Chemicals are used among other things to provide function in materials and products. Some chemicals do show hazard characteristics that are of great concern. Thus, there is a clear need for phase out actions of such hazardous chemicals that today are used in materials and products. But, we cannot phase out chemicals without replacing their functionality.

In the manufacturing of textiles large amounts of chemicals are used. A quantitative study of the consumption of chemicals during the life cycle of

C. Jönsson · S. Posner · S. Roos (✉)
Swerea IVF, Box 104, 431 22 Mölndal, Sweden
e-mail: Sandra.Roos@swerea.se

© Springer Nature Singapore Pte Ltd. 2018
S.S. Muthu (ed.), *Detox Fashion*, Textile Science and Clothing Technology,
DOI 10.1007/978-981-10-4876-0_1

1

textiles showed that between 1 and 5 kg of chemicals are used per kg textiles (Olsson et al. 2009). Some of the substances are harmful to health and/or the environment, with properties such as sensitizing, human toxic, eco-toxic, persistent or bio-accumulative (Munn 2011).

This chapter presents a model for practical substitution including also the technical and economical performance of alternative chemistries. The first section gives a background to why hazardous chemicals are used in textile production and occur in the ready-made textile product. Furthermore it describes the state-of-the-art regarding the legal and voluntary initiatives to phase-out hazardous chemicals in the textile industry. Finally the conditions required for a viable substitution are explained. Two specific examples are addressed where the authors have in practice applied, and iteratively developed, the suggested substitution model. The latter example involves textile chemistry and is a further development of a model used for phase out of hazardous flame retardants in plastic components in electronics as well as textiles.

1.1 Use of Chemicals in Textile Production

A wide variety of chemical substances with various functionalities, applications and properties are used in textile manufacturing. Chemical substances can be grouped in several different ways (Swedish Chemicals Agency 2004a), based on:

- chemical structure (phthalates, polychlorinated biphenyls etc.),
- functional properties (plasticizers, flame retardants etc.), or
- toxicological properties (endocrine disrupters, carcinogens, etc.).

While the chemical structure is a singular property, both the functional and toxicological properties are not; one substance may have one or many functional properties (e.g. both be a plasticizer and a flame retardant) and also one of many toxicological properties (e.g. both be endocrine disruptive and carcinogenic).

Figure 1 shows the long sequence of process steps in textile production and the type of chemicals that are used in each step. In this overview, the chemicals have been described after their functional properties. The grouping of chemicals after their functional properties is a key factor in the model for practical substitution. The functional properties of chemicals can be further divided into:

- **Effect chemicals**, which provide function to the final textile product (softeners, plasticizers etc.). These functions are usually selected by the product designer and/or the procurer. Sometimes this group of chemicals is addressed as "functional chemicals", hence giving function to the final product.
- **Processing chemicals**, which are used in the processing of textiles in the production (antifoaming agents, catalysts etc.). These functions are selected by the process engineer or sometimes specified by the chemical company to achieve compatibility with chemicals added to provide final effect.

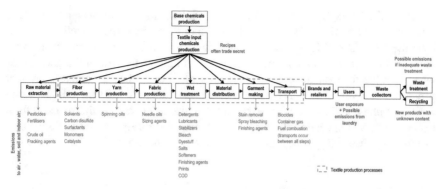

Fig. 1 Examples of commonly occurring chemicals (in *red*) in the textile life cycle. The chemicals are described after which function they deliver in each process step. Figure from Roos (2016)

This way of grouping the chemicals in two sets (process chemicals vs. effect chemicals) will facilitate for companies to organize and target their chemical management including the efforts to substitute hazardous chemicals (Roos 2016).

1.1.1 Effect Chemicals

Effect chemicals (or functional chemicals) are added to give an article a specific function. Effect chemicals contribute to design or any feasible technical function in the final product, e.g. colorants. Flame retardants in clothing and furniture save a large number of lives each year. Clothing with a high degree of water and dirt repellence is necessary in many workplaces but also in high demand in sports and for the comfort in everyday life. Further, we need to use biocides for a safe and reliable healthcare situation. However, these chemicals bearing toxic properties contribute to the environmental burden of this class of products. For effect chemicals there is a need for certain concentration in the final product in order to achieve the desirable function (Swedish Chemicals Agency 2014).

Some examples of functional (or effect) chemicals are:

- Colorants (dyestuffs and pigments)
- Oil, soil and water repellents
- Plasticizers
- Flame retardants
- Fragrances
- Alloys
- Biocides for defined functionalities in the article e.g. disinfectants
- Stabilizers e.g. antioxidants, UV/light stabilizers and anti-degradants.

The effect chemicals that are used should have good compatibility, such as good solubility in the materials (Posner 2009). Some effect chemicals require good affinity to the fibers, for example as dyes in cellulose. In order to sustain the desired function in the final textile product during the usage phase the "function" should have the most favorable ageing characteristics possible. In other words, there is no point in the functional chemistry to last twice as long as the rest of the garment or vice versa the for instance color to fade after the first washes for a textile product meant to last several years. The effect chemicals are not only relevant to high degree in the use phase and the possible exposure, but also to the end-of-life scenario for a textile product. Especially chemicals threatening the vision of non-toxic circular materials or disturbing a recycling process are of importance.

1.1.2 Processing Chemicals

The other category are called processing chemicals, also called auxiliary chemicals, that are necessary to make processes work, but they do not provide any desired properties to the final product and are therefore not meant to remain in the finished product.

Some examples of process chemicals are:

- Organic solvents
- Surfactants e.g. wetting and dispersing agents
- Softeners
- Curative agents
- Accelerators
- Chain extenders
- Lubricants
- Defoaming agent
- Catalysts
- Hardening agents
- Vulcanizing agent (rubber)
- Retarder (rubber)
- Complexing agent
- Salts
- Acids and bases
- Reactive resins (e.g. binders and adhesives) for various finishing treatments
- Biocides as preservatives in the process or during storage and transport e.g. fungicides and preservatives.
- Tanning agents (leather)
- Drying agents
- Intermediates, precursors and monomers.

Remains of the process chemicals may be found in the finished product and cause health and/or environmental problems. A process chemical which remains as impurities in the final product often has a relatively low concentration, compared with the concentration of an effect/functional chemical in the final product.

1.2 Development of Legal and Voluntary Restrictions of Chemicals Content in Textile Products

In 2002, when the global leaders gathered at the Johannesburg World Summit on Sustainable Development, detrimental impacts from chemicals was one of the highlighted sustainability challenges. The participating countries agreed to the following goal:

> by the year 2020, chemicals are produced and used in ways that minimize significant adverse impacts on the environment and human health
>
> (United Nations 2002, paragraph 23)

This goal was then specified further at the International Conference on Chemicals Management (ICCM) in Dubai four years later, where the Strategic Approach to International Chemicals Management (SAICM) was adopted (UNEP 2006). This policy framework for sound management of chemicals has since then affected chemicals management in several sectors including textile and fashion.

1.2.1 Development of Legal Restrictions Addressing Chemicals in Textiles

In 2007, the European chemicals legislation REACH (Registration, Evaluation, Authorisation and Restriction of Chemicals) (European Commission 2006) entered into force. The REACH legislation was a harmonization of the chemicals legislation in the European Union and the European Economic Area (EEA) countries (Norway, Iceland and Lichtenstein) (hereinafter called the EU). REACH also implied that chemicals became regulated in many product groups where chemicals were not previously regulated in the EU. Another effect that REACH brought about was that several other countries followed the REACH example and developed similar regulations, in popular terms often called "China REACH" (China Ministry of Environmental Protection (MEP) 2010), "India REACH" (Government of India 2012) or "K-REACH" (South Korean Ministry of the Environment 2011).

However, the possibilities to use legislative measures as a tool to counteract problems with hazardous chemicals in textiles is impaired by a major challenge: the fact that any national (or federal) regulation of chemicals is limited to actions inside their area of jurisdiction (Roos 2015). This limitation means that, for example, the European legislation can only regulate the chemical content of products produced in, imported to or used in the EU. The textile supply chain is on the contrary global to its nature; textile products and semi-finished products are constantly exported and imported across country borders. In the absence of a legislative framework that covers the entire textile supply chain, the industry has instead acted via voluntary initiatives to secure a responsible chemicals management. Legislation can thus be identified as one measure to achieve the SAICM objectives, where industry voluntary action is another of at least equal importance (Roos 2015).

1.2.2 Development of Voluntary Restrictions Addressing Chemicals in Textiles

Today, there are many on-going international activities for development of voluntary schemes addressing hazardous chemicals in textiles. For example, a broad range of textile labels exist; currently 108 textile ecolabels are listed in the Ecolabel Index (Ecolabel Index 2016). The most common environmental textile label globally is the Oeko-Tex® 100 certification (OEKO-TEX® Association 2017), followed by BlueSign (BLUESIGN® 2017) and Global Organic Textile Standard (GOTS 2017). One important recent initiative is the Outdoor Industry Association (OIA) and the Sustainable Apparel Coalition (SAC) who jointly developed the Chemicals Management Framework, also called the Chemicals Management Module (CMM) (OIA 2014) which has also been integrated into SAC's Higg Index 2.0 (SAC 2017). Several other management tools addressing specific chemicals are available through the Substitution and Alternatives Assessment Toolbox (SAAT), recently developed by the Organisation for Economic Co-operation and Development (OECD) (OECD 2015).

In order to show compliance with existing legislation and to fulfill customer demands, some kind of chemical management is utilized by all actors on the textile market. Current company chemical management systems are designed and intended to handle the obstacles. For textile brands many companies use so called Restricted Substance List (RSL) as their core tool in the chemical management system. The RSL normally consists of chemical names as well as CAS numbers[1] of the specific substances (AAFA 2015). In some cases the lists also give guidance to where the chemicals may be found in the production or with information of the function provided (AFIRM 2015; Swedish Chemicals Group/Swedish Textile Importer's Association 2016). Substances included in such lists mainly consist of restricted substances often but not always related to the textile materials (Roos et al. 2017). Some front runners occasionally and additionally include substances that may be, but not yet are, regulated in their RSLs together with already regulated and often textile-relevant substances such as carcinogens, endocrine disruptors and skin sensitizers (ChemSec 2017).

Most of the above mentioned tools has in common that they provide "negative lists", that is lists of unwanted substances, and little guidance on how to perform the substitution. The exception is BlueSign and GOTS that also provide "positive lists" of chemical products that have been evaluated and identified by the schemes as more environmentally friendly alternatives.

---

[1]Chemical Abstracts Service Registry Numbers.

1.3 Substitution of Substances, Materials or Products

Most of the work with substitution of hazardous chemicals aims at finding a substitute substance with less harmful properties than the original substance (Swedish Chemicals Agency 2007). In practice, substitution can refer either to chemical substances, to chemical products (i.e. commercial mixtures), to materials and even to products (Leonards 2011). Furthermore, the approach for substitution can be based on either hazard or risk. Inherent safety, also called primary prevention, consists in the elimination of a hazard. It is contrasted with secondary prevention that consists in reducing the risk associated with a hazard (Swedish Chemicals Agency 2007). Regardless of approach and type of substitution, the substitute product system (where the substitute can be a chemical substance, chemical product, material, technical solution or textile product) need to provide the same function as was provided in the original product system.

In the textile industry, a lot of the focus has been put on effect chemicals with hazardous properties, such as halogenated flame retardants and perfluorinated durable water repellent (DWR) agents. The requirements on alternative effect chemicals are therefore described below.

1.3.1 Requirements on Alternative Effect Chemicals

Substitute effect chemicals can appreciably impair the properties of the textile material. The basic problem in substitution is the trade-off between the decrease in performance of the textile material caused by the substitution and the (perceived) customer requirements and the foreseen lowered hazard. In addition to fulfilling the appropriate function (color, fire retardants, water repellency etc.), a viable alternative effect chemical shall, at most, fulfill all of the below qualities (Norwegian Pollution Control Authority (SFT) 2009):

- Functional properties

  - Provide the same function as was provided by the original effect chemical
  - Not alter electrical properties

- Mechanical properties

  - Not significantly alter the mechanical properties of the textile material
  - Be easy to incorporate into the host textile material
  - Be compatible with the host textile material
  - Be easy to extract/remove for recyclability of the textile material

- Physical properties

  - Be colorless or at least non-discoloring (not applicable for dyestuffs and pigments)
  - Have good light stability

- Be resistant towards ageing and hydrolysis
- Not cause corrosion

- Health and environmental properties

  - Not have harmful health effects
  - Not have harmful environmental properties

- Commercial viability

  - Be commercially available and cost-effective.

1.3.2 State-of-the-Art Substitution Work

Among the on-going initiatives with phasing-out hazardous chemicals in the textile industry, the focus is mainly on evaluating the environmental and health aspects of chemicals, and describing the current emission levels. The technical performance in the end application for the possible substitute chemicals does not receive the same attention (AAFA 2015; ChemSec 2017; UNEP/POPS/POPRC.8/INF/17 2012; ZDHC 2014). In addition, many research projects committed to evaluating hazardous substances and their possible alternatives also neglects the technical performance (Howard and Muir 2010; Quinete et al. 2010; Wang et al. 2013a). The technical performance is left to the companies to evaluate. This may lead to inefficiency in the substitution process and also have the consequence that companies never dare to take the step to practical substitution, at least not in a proactive way.

Further, focus is sometimes put on equal technical performance but also the performance level can be questioned. That in turn requires performance criteria to match the new performance level. But also the "culture" and the way things have previously been working or performed need to be questioned. In other words the substitution work needs much more attention on the performance side.

Many substitution processes result in solutions that are rather specific for specific applications compared to some of the former solutions. Some examples are halogenated flame retardants used in a variety of applications that needs a several substitutes to cover the same number of applications.

1.4 Life Cycle Perspective

Life cycle assessment (LCA) is an ISO-standardized method for quantitative evaluation of the environmental performance of products and services throughout the life cycle (Baumann and Tillman 2004). The life cycle of a product is generally divided into four main phases: raw material extraction, production, use and end-of-life, see Fig. 2. Environmental performance is measured in the form of environmental impacts of emissions to air, water and soil as well as consumption of resources (energy, water, land and material), in the different stages of the life cycle.

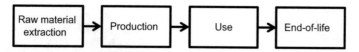

Fig. 2 The four life cycle phases of a product or service

The life cycle perspective of the LCA is essential in order to avoid sub-optimization, i.e. improving just a part of a system in a manner that negatively affects other parts of the system. Sub-optimization can occur when only parts of the life cycle are studied and the overall performance is not evaluated. The life cycle perspective is thus important also for substitution work, to assure that the substitution of a substance in one sub-part of the life cycle (e.g. application of a durable water repellent coating) does not lead to increased toxicity in another sub-part (e.g. increased toxic emissions from dyehouses due to a shorter life length of the product).

2 Model Creation

This section gives the background to the creation of the model and the context in which it was developed. Experts in the fields of environmental, analytical, organic and physical chemistry, physics, ecotoxicology, economics, risk assessment and life cycle assessment (LCA) were all involved and engaged in order to take a holistic view on substitution of substances. Further, the requirements for a viable substitution in practice have been identified in dialogue with industry partners, which has been crucial for understanding the implications of the real-life changes that need to be made in the supply chains.

The development of the model for practical substitution has been based on empirical experiences from case studies. In systems analysis, case studies play an important role for method development where, according to Dubois and Gadde (2002), using a logic-based systematic combining approach "has been showed to be particularly useful for the development of new theories", letting methodological framework, empirical fieldwork, and case analysis evolve simultaneously.

2.1 First Development Steps

The creation of the model suggested here began in the project ENFIRO (Life Cycle Assessment of Environment-Compatible Flame Retardants).[2] ENFIRO followed a prototypical case study approach in which existing and alternative flame retardants were evaluated regarding their flame retardant properties, their influence on the

---

[2]http://www.enfiro.eu/, http://cordis.europa.eu/result/rcn/56829_en.html.

Fig. 3 First version of the
model: the ENFIRO chemical
alternative cycle

function of products once incorporated, and their environmental and toxicological properties.

There were at the time several non-brominated flame retardants existing on the market. However, there was limited information available about the environmental and toxicological impact of these alternatives. Furthermore, the alternatives were difficult to apply before tests had shown that they did not adversely affect the quality of consumer products. ENFIRO evaluated viable substitution options for a number of brominated flame retardants. These flame retardants were studied in five applications: textile coatings, intumescent paint, electronic components, printed circuit boards and injection moulded products. In Sect. 3.2, the results from the case study with textile coatings are provided.

The ENFIRO approach developed followed a chemical substitution cycle anchored in four major elements (Fig. 3). In the first element the alternative halogen-free flame retardants were prioritized and the most viable alternatives selected. The second major element focussed on the technical performance (fire and application), hazard and exposure assessment of the selected halogen-free flame retardants. The collected information was analyzed in a comparative hazard and risk assessments (third element). Finally, information on production costs and socio-economics of the halogen-free flame retardants and related products, was added to give a holistic picture together with impact assessment studies using life-cycle assessments (LCA) (fourth element). This finally resulted in a recommendation of certain halogen-free flame retardants and their related product combinations.

2.2 Generalization of the Model

The ENFIRO model was built to meet the needs of the specific needs of the ENFIRO project. In order to develop procedures that can be used both as a basis for

Fig. 4 The generalized
model for practical
substitution

legislation and for substitution in practical product development, in textiles as well
as other consumer products, a generalization of the model was needed.

The generalized model is intended to be able to be applied on any substance
group and to any type of substitution (the substitute can be a chemical substance,
chemical product, material or product). In addition, the model should be able to be
used by any consortia covering the fields of environmental, analytical, organic and
physical chemistry, physics, ecotoxicology, economics, risk assessment and life
cycle assessment (LCA) that is needed for the specific case in order to achieve
robust results. Which these fields are for a specific case need to be identified in
dialogue with academic and industry partners.

To the model was added a first step where mapping and characterization of
substances of high concern are related to other substances of concern as a way to
prioritize chemicals and products to be included in substitution processes (Fig. 4).
The next steps are based on the ENFIRO approach for substitution of hazardous
substances, including testing technical and economical viability.

3   Evaluation via Case Studies

The model for practical substitution of hazardous chemicals has been evaluated in
several case studies, whereof two in the textile field: water and soil repellent textile
coating materials (the SUPFES project[3]) and flame retarded textiles (the ENFIRO
project[4]).

---

[3]http://supfes.eu/.

[4]http://www.enfiro.eu/, http://cordis.europa.eu/result/rcn/56829_en.html.

3.1   The SUPFES Project

The project SUPFES (Substitution in practice of prioritised fluorinated compounds for textile applications),[5] 2014–2017, has taken a holistic view on the use of per- and polyfluoroalkyl substances (PFAS) in the textile industry studying emissions, life cycles as well as human and aquatic toxicity. To create possibilities for real change the SUPFES project also looks at practical substitution of fluorinated compounds in textile applications. This means that substitutes need to be evaluated not only for their health and environmental properties but they need also to be evaluated for their technical properties (function, durability and compatibility with textile processing). Within this project the level of performance have also been thoroughly discussed for different textile applications since the performance level provided by perfluorinated compounds are high but rather many alternatives exist for lower levels of performance.

3.1.1   Current Use of Per- and Polyfluoroalkyl Substances (PFAS)

Per- and polyfluorinated chemical products are extremely versatile and are used in a variety of industrial and consumer applications and products. Some of these chemical products contain or release per- and polyfluoroalkyl substances (PFAS) that are documented to be persistent as well as accumulating in the human body as well as in the environment. The most common textile applications for PFAS-based chemicals are as water- and oil-repellent agents (for so called durable water repellent (DWR) treatments). Other applications include use in digital printing processes to prevent bleed-through of the fabric.

The PFAS substance group is known to be (or transform into substances that are) persistent, i.e. does not degrade in the environment. PFAS are in addition bioac-cumulative, i.e. their concentration in organisms can become higher than that of the surrounding environment. PFAS have been detected in the ground and water in remote areas, such as the Arctic as well as in the blood of small children, adults and other mammals (Posner et al. 2013). The rising levels of PFAS found in the environment are of high concern because these substances have been linked to adverse health effects, such as delayed puberty onset, elevated cholesterol levels, reduced immunologic responses to vaccination and over-representation of attention-deficit/hyperactivity disorder (ADHD) in children (Bergman et al. 2013). Two PFAS are currently subject to legal restriction: perfluorooctane sulfonic acid (PFOS) with CAS RN 1763-23-1 is restricted under the global Stockholm Convention and perfluorooctanoic acid (PFOA) with CAS RN 335-67-1 is restricted in Norway. Other PFAS are proposed for regulations.

---

[5]http://supfes.eu/.

The fluorochemistry has been in production and on the global market since the early 1950th where the so called long-chain fluorochemistry[6] has dominated the market for decades. There are two main chemistries that used to dominate this market namely the so called PFOS-related chemistry and the PFOA-related chemistry. In 2003 there was a voluntarily phase out in 2003 of the production of PFOS by the most important global producer 3 M, which marked a major decrease in global production and use. Production before 2003 was mostly for surface treatment such as textile and for paper protection (UNECE 2006). Subsequently several countries and regions worldwide have introduced phase out programs and legislation to limit the use of the PFOS-related chemistry.

In 2009 PFOS and related substances where declared Persistent Organic Pollutants (POP) by the Stockholm Convention that is one of the UNEP activities under the SAICM strategy. For a long period PFOA-related chemistry was considered an alternative to PFOS-related chemistry, but this too has demonstrated severe health and environmental properties and is currently in several global phase out programs to be replaced by the so called short-chain fluorochemistry and non-fluorinated chemistries such as polysiloxanes, waxes and paraffins.

3.1.2 Characterization of Original Substance in Use

The situation with PFOS and PFOA being restricted in several applications at the same time as the market is searching intensively after alternative chemistries has led to that manufacturers of DWR agents are very secretive with the content of their products. The first step of the SUPFES project was therefore to characterize the PFAS currently in use.

Characterization of PFAS in use

Textile materials on the market were screened for the traditional long-chain PFAS (hereafter called old PFAS), alternative PFAS, and novel fluorinated compounds. The main aims were to characterize the diffuse sources and identify the PFAS used in materials and goods. A literature search and the SUPFES stakeholders were used to associate different types of PFAS with different chemical products. This information was used to collect materials and goods, followed by an analytical screening of PFAS.

---

[6]The term "long-chain PFAS" has been defined by OECD (2013) as:

  i. PFCAs with 7 and more perfluoroalkyl carbons, such as PFOA (with 8 carbons or C8 PFCA) and PFNA (with 9 carbons or C9 PFCA);
  ii. PFSAs with 6 and more perfluoroalkyl carbons, such as PFHxS (with 6 perfluoroalkyl carbons, or C6 PFSA) and PFOS (with 8 perfluoroalkyl carbons or C8 PFSA); and
  iii. Substances that have the potential to degrade to long-chain PFCAs or PFSAs, i.e. precursors such as PASF- and fluorotelomer-based compounds.

Table 1 List of PFASs analyzed in textile samples

Compound	Abbreviation	CAS. nr.	Formula
Perfluorobutanoic acid	*PFBA*	375-22-4	C_3F_7COOH
Perfluoropentanoic acid	*PFPeA*	2706-90-3	C_4F_9COOH
Perfluorohexanoic acid	*PFHxA*	307-24-4	$C_5F_{11}COOH$
Perfluoroheptanoic acid	*PFHpA*	375-85-9	$C_6F_{13}COOH$
Perfluorooctanoic acid	*PFOA*	335-67-1	$C_7F_{15}COOH$
Perfluorononanoic acid	*PFNA*	375-95-1	$C_8F_{17}COOH$
Perfluorodecanoic acid	*PFDA*	335-76-2	$C_9F_{19}COOH$
Perfluoroundecanoic acid	*PFUnDA*	2058-94-8	$C_{10}F_{21}COOH$
Perfluorododecanoic acid	PFDoDA	307-55-1	$C_{11}F_{23}COOH$
Perfluorotridecanoic acid	PFTrDA	72629-94-8	$C_{12}F_{25}COOH$
Perfluorotetradecanoic acid	PFTeDA	376-06-7	$C_{13}F_{27}COOH$
Perfluorobutane sulfonate anion	PFBS	45187-15-3	$C_4F_9SO_3^-$
Perfluorohexane sulfonate anion	PFHxS	108427-53-8	$C_6F_{13}SO_3^-$
Perfluoroheptane sulfonate anion	PFHpS	375-92-8	$C_7F_{15}SO_3^-$
Perfluorooctane sulfonate anion	PFOS	45298-90-6	$C_8F_{17}SO_3^-$
Perfluorooctane sulfonamide	FOSA	754-91-6	$C_8F_{17}SO_2NH_2$
6:2 Fluorotelomer sulfonate anion	6:2 FTSA	425670-75-3	$C_6F_{13}CH_2CH_2SO_3^-$

Textile samples from outdoor clothing like trousers, jackets, gloves etc., were provided by different suppliers from the outdoor industry. A first screening of the materials for the content of the traditional long-chain PFASs and alternative short chain PFASs (Table 1) was performed by a simple and at the time non-validated extraction method with methanol as extraction solvent and quantification by LC-MS/MS. Results of the screening showed that a variety of PFASs were present in the samples, with some PFOS and PFOA concentrations exceeding the norm set by the European Union for PFOS (1 $\mu g/m^2$) (EU 2006) and the Norwegian government for PFOA (1 $\mu g/m^2$) (Lovdata 2014). PFASs profiles showed that for some samples comparability patterns were found. No comparability was detected between the type of chemistry used by the textile industry (C_6 or C_8) and the profiles detected by the screening (Andersson et al. 2014).

Development of an extraction method of PFCAs, PFSAs and FOSA in textile samples

Although some analyses on PFASs in textiles have already been performed by others (Herzke et al. 2012; Knepper et al. 2014), no peer reviewed validated extraction methods were published. For the screening of PFASs in the 34 textile samples methanol was used, because it had already successfully been used for extracting PFASs from several matrixes (van Leeuwen et al. 2009; Backe et al. 2013; Weiss et al. 2013; Wang et al. 2013b). However, because Knepper et al. (2014) used acetone/ acetonitrile (80:20, v/v) to extract PFASs from textile samples, both the acetone/acetonitrile mixture as well as methanol were optimized as extraction solvent. The number of sequential extractions and extraction time on a shaking device were optimized by performing sequential extractions of five textile samples with either acetone/acetonitrile (80:20, v/v) or methanol and with different extraction times.

The developed extraction method (Van Der Veen et al. 2016) for PFASs analysis in textile samples is validated by a recovery assessment, a repeatability assessment, and a reproducibility assessment. For the recovery assessment two textile samples were spiked in triplicate on two different concentration levels. For the repeatability assessment three samples were extracted in triplicate in the same series. For the reproducibility assessment the three samples were extracted in triplicate on three different days.

The 34 textile samples have been reanalyzed for, together with 11 additional textile samples, with the developed and validated method. Five of the 11 additional samples consisted of two different colors. Those different colors were analyzed as separate samples, which resulted in a total of 50 textile samples. Although 4:2 FTSA, 6:2 FTSA and 8:2 FTSA were not included in the validation, those PFASs were quantified with the developed method as well. PFASs profiles were calculated for the textile samples based on the results.

Environmental and exposure assessment of PFAS

For the environmental and exposure assessment, analytical methods for the outdoor environment were developed. Physical-chemical property data for the identified PFAS were taken from the literature or generated using existing structure-property relationship models (EPIsuite, SPARC). The property data were used to predict the environmental fate of the chemicals. Key matrices for environmental and human exposure (sewage sludge, effluent, food, air) were sampled and analyzed for the chemical(s). Waste water and sewage sludge were sampled in a nested design to identify variability in contaminant flows in collaboration with a wastewater treatment plant in Stockholm, Sweden. Additionally, key performance data for sludge management at the wastewater treatment plant were collected to facilitate system analysis of the consequences of increased and decreased sludge quality on the environmental performance of sludge management. To assess indoor contamination, dust and indoor air samples were analyzed from microenvironments where textile consumer products are used (homes, offices, clothes shop). The analytical data were used to calculate environmental and human exposure to the identified PFAS.

Selection of prioritized substances

The prioritization was made using a scheme developed for the purpose, where all elements of the analyses above were set in preference order. The scheme had five elements: (1) sources, (2) product/PFAS combinations, (3) leaching and emission, (4) environmental levels, and (5) toxicological and ecotoxicological properties of the PFAS. The prioritized results (Table 2) served as the basis for the selection of alternative compounds.

Table 2 Prioritized PFAS to be substituted

Substance	Abbreviation	CAS no	Occurence
Long chain fluoro chemistry—Reference chemistry in SUPFES			
Perfluorooctanoic acid	PFOA	335-67-1	Stable degradation product
8:2 fluorotelomeralcohol	8:2 FTOH	678-39-7	Precursor
Short chain fluorochemistry			
Perfluorohexanoic acid	PFHxA	307-24-4	Stable degradation product
6:2 fluorotelomeralcohol	6:2 FTOH	647-42-7	Precursor
Perfluorobutanesulfonic acid	PFBS	45187-15-3	Stable degradation product
Perfluorobutanoic sulfonamidoalcohols	FOSE		Precursor

3.1.3 Alternative Selection

A list of potential and already implemented alternatives was first assembled. Both PFASs with shorter chain lengths than the old PFASs (e.g. based on perfluorobutanesulfonic acid (PFBS) or 6:2 instead of 8:2 fluorotelomer chemistry) as well as non-fluorinated alternatives (siloxanes/polysiloxanes, dendrimers and waxes). The selection of potential alternatives was based on the "Technical paper on the identification and assessment of alternatives to the use of perfluorooctane sulfonic acid in open applications" (UNEP/POPS/POPRC.8/INF/17 2012) and on information from the associated partners. A literature review was prepared in which a detailed summary of the different chemistries in durable water repellency (DWR) products was provided (Holmquist et al. 2016). A range of common brands providing alternatives to long chain PFAS are listed in Table 3 (UNEP/POPS/POPRC.8/INF/17 2012).

Figure 5 shows a summary of the groups of chemicals that were selected as alternatives, namely hydrocarbons (including waxes) (HC-DWR), silicon based chemistry (Si-DWR) and short chain fluorocarbons (FC-DWR). In addition information about the repellency performance in relation to the end groups of each chemical structure is presented.

3.1.4 Toxicity and Exposure Assessment

The same procedure as described in Sect. 3.1.2 was applied to assess the identified alternatives that were already in use (e.g. PFBS-based chemicals). For the potential alternatives that were still in the development phase and not in commerce yet, a modeling approach was taken. A preliminary hazard assessment was undertaken of different DWR chemistries (Holmquist et al. 2016). This was then further refined as new information was created in experimental testing made in SUPFES and outside the project.

Table 3 A selection of alternatives to the use of long-chain PFAS for carpets, leather and apparel, textiles and upholstery

Chemical content	Product name	Brand	Application
Fluorine-free products			
Hyperbranched hydrophobic polymers (dendritic i.e. highly branched polymers) and specifically adjusted comb polymers as active components. Glycols are added as solvents and cationic surfactants in small amounts act as emulsifiers	RUCO-DRY ECO	Rudolf GmbH (Germany)	Superhydrophobic surfaces, meaning.contact angles larger than 150°. Rudolf Chemie describes the coating as a bionic Lotus coating addressed after the Lotus plant leaves. Applied in coatings, textile and leather
Siloxanes and silicone polymers Mixtures of silicones and stearamidomethyl pyridine chloride, sometimes together with carbamide (urea) and melamine resins	Advantex™	Bluestar Silicones	Impregnation of all-weather textiles Surfactants for the impregnation of textile fabrics, leather, carpets, rugs and upholstery and similar articles
Fluor-based alternatives			
Perfluorobutane sulfonate (PFBS) derivatives or other alternatives based on various C₄-perfluorocompounds Fluorotelomer alcohols and esters	Scotchgard TM Zonyl® Capstone®	3 M Du Pont	Applied in coatings, printing, and textiles
Fluorinated polymers	Foraperle® 225, etc.	Du Pont	Impregnation of leather and indoor car upholstery

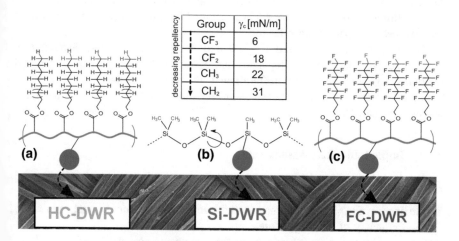

Fig. 5 Groups of chemicals: hydrocarbons (including waxes) (HC-DWR), silicon based chemistry (Si-DWR) and short chain fluorocarbons (FC-DWR) including their respectively repellency performance. Figure modified from Holmquist et al. (2016)

Physical-chemical properties were estimated using structure-property relationship models (EPIsuite, SPARC). Different emission scenarios were used in the model and the alternatives as well as the old PFASs (benchmarks) were computed in the estimation model. Conclusions could then be drawn in terms of the probable behavior, transport and fate of the alternatives relative to the old PFASs (i.e. are the alternatives "better" or "worse").

A literature search for toxicological and ecotoxicological properties of selected alternatives was carried out and the data was compared to corresponding data from the old PFASs. In vitro toxicity testing was performed both for the alternatives as well as for the old PFASs. The established in vitro assays included testing for induction of endocrine disruption, dioxin-like toxicity, genotoxicity, cytotoxicity, and thryroid hormone binding. The in vitro toxicity was further tested after a biotransformation step using different species' microsomes. Additionally, two standard subchronic ecotoxicity tests were performed with brackish/marine species, i.e. a macroalga (*Ceramium Tenuicorne*) and a crustacean (*Nitocra Spinipes*) representative of the Baltic Sea. In all toxicity and ecotoxicity testing, the old PFASs were used as benchmark chemicals, with the aim to characterize the relative toxicity of the alternatives ("better" or "worse" approach).

3.1.5 Case Studies

The overall aim of SUPFES is to demonstrate practical substitution in consumer products. The textile industry sector had clearly expressed needs of meeting both legal and customer demands for surface treatments aiming at water and dirt repellence. To ensure the success of the proposed substitution model, case studies using alternatives in prototypes were performed using a life cycle perspective. Different textiles were selected to represent important sources studied in Step 1 of the substitution model. The case studies included both prototype manufacturing as well as technical performance testing.

The Swedish outdoor brand Haglöfs and the research institute Swerea IVF commissioned water and dirt repellent fabrics for outdoor jacket using proposed alternatives (fluorinated and non-fluorinated alternatives) using conventional solvent phase chemistry and gas phase chemistry (plasma), for input to the technical testing and the emission studies.

3.1.6 Impact and Risk Assessment

Risk estimates for prototypes were based on scenarios describing

1. the degree of exposure of humans and the environment under normal conditions,
2. the frequency of these exposures,
3. the probability for the occurrence of these exposure scenarios, and
4. the hazard characteristics of the substances in combination.

A newly developed methodology by one of the project partners, Vrije Universiteit (VU) in Amsterdam, for measuring emissions of chemicals from materials to air under controlled conditions was used for non-targeted as well as targeted analysis. This methodology enables analysis of chemical compounds in materials; test for emissions to air, analyze in air, dust, wastewater streams, sewage sludge, and recipient water as well as take water samples from a washing machine.

Life cycle assessment (LCA) was used to evaluate the balance of inputs and outputs of systems in different categories related to resource use, human health and ecological areas in different steps. Information on potential toxic effects on human health and the environment, along with physical and chemical properties of the alternatives of interest, was used as input to the LCA. The scenarios in the LCA examined a base case (current products), an improved case (with substitutions) and a zero option (where such substances are not used and thus the technical performance is not provided). The impacts on downstream processes (e.g. wastewater treatment and sludge management) were considered for these scenarios.

To verify the economic viability of suggested designs, life cycle cost (LCC) calculations were performed. The economic viability of the developed technology will be analyzed and interpreted with the chosen software.

3.1.7 Technical Performance

Products with PFASs are often made for outdoor use due to their water and oil repelling and release properties. In order to validate the performance of the prototype(s) developed in the project and to compare the prototype(s) with the existing fluoro-based technology, the following properties were taken into account:

- General properties; washability, compatibility with dyestuffs
- Mechanical properties; resistance to abrasion and tearing
- Physical properties; water vapor resistance ("Skin Model"), water and oil repellency, overall comfort

Durable water repellence (DWR) and related performance attributes include not just water repellency, but normally performance for below attributes is required:

- Water repellency: various static and dynamic tests are used to measure water repellency. Since there is no set definition of water repellency, the conditions of the test must be stated when specifying water repellency. Water repellent fabrics are generally defined as fabrics which resist being wetted by water; water drops will roll off the fabric. Water repellency depends on the nature of the fiber surface, the porosity of the fabric and the dynamic force behind the impacting water spray.
- Water proof: this concept is in itself an overstatement; a more descriptive term is "impermeable to water".

- Oil repellency: tested by placing a drop of oil on the fabric and observing whether the drop resides on top the fabric or whether it penetrates. A homologous series of hydrocarbons decreasing in surface tension is used to rate the fabric's oil repellency. The hydrocarbon with the lowest surface tension to remain on top and not penetrate is indicative of the fabric's repellency. The lower the surface tension of the liquid, the better is the fabric's resistance to oily stains.
- Stain repellency: the ability of a treated fabric to withstand penetration of liquid soils under static conditions involving only the weight of the drop and capillary forces.
- Durability (e.g., to laundering, light exposure, abrasion, dry cleaning, etc.).
- Other (Soil repellency, stain release, soil release etc.).

It is important to distinguish between water repellent and water proof fabrics. A fabric is made water repellent by depositing a hydrophobic material on the fiber's surface; however, waterproofing requires filling the pores as well. Water repellent fabrics have open pores and are permeable to air and water vapor, hence they provide the function of "breathability", i.e. moisture can be transferred through the material. Water repellent fabrics will always permit the passage of liquid water once hydrostatic pressure is high. Water proof fabrics on the other hand are resistant to the penetration of water under much higher hydrostatic pressure than water repellent fabrics, but do not provide the breathability function.

3.1.8 Finalizing the SUPFES Project

The SUPFES project will be finalized during 2017.

3.2 The ENFIRO Project

The project ENFIRO (Life Cycle Assessment of Environment-Compatible Flame Retardants),[7] 2009–2012, investigated the substitution options for some brominated flame retardants and compared the hazard, exposure, fire, and application performances.

Many brominated flame retardants are known to have unintended detrimental effects on the environment and human health. The situation before ENFIRO started was that less toxic alternatives appeared to already be available on the market, however, comprehensive information on their possible toxicological effects was lacking. ENFIRO investigated the substitution options for some brominated flame retardants and compared the hazard, exposure, fire, and application performances. Based on these results, risk and impact assessments were carried out. In total 14

---

[7]http://www.enfiro.eu/, http://cordis.europa.eu/result/rcn/56829_en.html.

halogen-free flame retardants as alternatives for decabromodiphenyl ether (decaBDE), tetrabromobisphenol A (TBBPA), and brominated polystyrenes (BPS) were selected. These flame retardants were studied in five applications: textile coatings, intumescent paint, electronic components, printed circuit boards and injection moulded products.

ENFIRO followed a prototypical case study approach in which new alternative flame retardants were evaluated. The evaluation included flame retardant properties, their influence on the function of products once incorporated, and their environmental and toxicological properties. The main objectives were:

- To deliver a comprehensive dataset on viability of production and application, environmental safety, and a life cycle assessment of the alternative flame retardants.
- To recommend certain flame retardant/product combinations for future study based on risk and impact assessment studies.

3.2.1 Current Use of Flame Retardants

Fires are among the most common causes of harm to people and property around the world. In the last ten to fifteen years, a number of risk-benefit analyses have been performed based on cases with actual fires. The conclusion from these risk-benefit analyses is that measures taken to improve fire safety lead to a clear reduction in the number of deaths and severe injuries. Such measures include increased and improved applications of various forms of flame-retardants. The need for improved fire safety has stimulated the development of better and more effective flame retardants. Also the development of legislation and extensive safety requirements for protection against fire has given rise to tough fire standards for a number of materials handled in situations where there is a risk of fire (Swedish Chemicals Agency 2006, 2004b).

Some halogenated flame retardants have unintended detrimental effects on the environment and human health, and are subject to legal restrictions. The European directive 2011/65/EU on the restriction of the use of certain hazardous substances in electrical and electronic equipment (RoHS) (European Commission 2011), addresses for example several halogenated flame retardants while for textile applications, flame retardants are regulated by the European Regulation (EC) No. 1907/2006 REACH (Registration, Evaluation, Authorisation and Restriction of Chemicals) (European Commission 2006).

Today, flame retardants are mostly used in the area of electronics, for example in the manufacturing of printed circuit boards, plastic casings for electronics, including mobile phone equipment. Polymeric materials are by far the most common material type containing flame retardants; the largest quantities of flame retardants (around 90%) are supplied to raw-material manufacturers in the plastics industry. A smaller proportion of world production of flame retardants (around 10%) is supplied to the textile and paper industries. The aromatic polybrominated diphenyl ethers (PBDE) have been one of the most commonly used flame retardants. The compound containing ten bromine atoms under the name decabromodiphenyl ether (decaBDE), with CAS RN 1163-19-5 is used in textile applications. DecaBDE is the most commonly occuring flame retardant among the organic aromatic bromine compounds. DecaBDE, which belongs to the category of additive[8] flame retardants, is produced in quantities of tonnes per year around the world. Furthermore, DecaBDE is always used in conjunction with antimony trioxide (ATO), which has a harmonized classification under the European Union CLP regulation as suspected of causing cancer (H351) (European Commission 2008).The EU Risk Assessment Report (RAR) of 2002 concluded that further information was required about the persistent, bioaccumulative and toxic (PBT) properties, and that DecaBDE is likely to be very persistent (vP). The substance has some similarities in its behavior to a very persistent and very bioaccumulative (vPvB) substance and its breakdown to substances with PBT or vPvB properties could occur in the environment (for example by metabolism in fish). The RAR further concluded that there are uncertainties regarding possible neurotoxic effects by mammals in laboratory studies as well as secondary poisoning. Possible formation of more toxic and accumulative products such as lower BDE congeners and brominated dibenzofurans in the environment should also be investigated.

The current knowledge about DecaBDE shows that the available assessment methodology might not be applicable to this substance, and in general also not to other brominated flame retardant's. It can be concluded that there is a continued need to monitor environmental contamination for both the substance in itself and also its more toxic and bioaccumulative degradation products. This uncertainty surrounding DecaBDE is expected to be clarified through further testing and long term biomonitoring (European Chemicals Bureau 2007).

3.2.2 Prioritization and Selection

In the first phase of ENFIRO a prioritization and selection of alternative flame retardants was carried out. The main objective was to identify a range of non-brominated flame retardants that were considered viable alternatives to specific commercial brominated flame retardants on the market. The identification was carried out using the scientific literature and other reliable scientific sources based on how they affect the material's characteristics of the polymers that are flame retarded. Such characteristics included compatibility, electrical properties, and various ageing properties and was based on already available data on toxicity, exposure risks and environmental fate. This resulted in the assessment of viability criteria for specific flame retardant applications that consisted of flame retarded marketable polymers.

---

[8]Additive means that the flame retardant is only physically bound to the flame retardant material, unlike the reactive flame retardants that are chemically bound.

At the start of the project an overview of existing data on alternative flame retardants was made. One of the most important findings was that large data gaps and contradicting information still existed for alternative flame retardants, which also showed the need for ENFIRO. The combination of polymers with the halogen-free flame retardants that were identified as commercially viable alternatives to specific commercial brominated flame retardants [TBBP-A, decaBDE, brominated polystyrene (BPS)] are presented in Table 4. The selection criteria were that the flame retardants should be halogen-free, commercially available, and that some information on the compatibility behavior in polymer materials should be available. The list of halogen-free flame retardants was further updated after consultation with the ENFIRO Stakeholder Forum consisting of flame retardant producers, formulators, end-users, environmental organisations, and others, and after initial screening tests. The list contains phosphorus flame retardants, inorganic tin-based flame retardants, nanoclays and combination of nanoclays with phosphinates. Based on the selected halogen-free flame retardants a literature survey of fire behavior including general characteristics of flame retardant chemicals, thermal degradation properties of the selected flame retardants, and a literature survey on the flammability and toxicity of the selected prototype base polymers and flame retardants was made. Literature data on the flame retardancy of selected systems using halogen-free flame retardants with comparison to brominated flame retardants were also presented.

One of the objectives of ENFIRO was to perform an ecotoxicological and health hazard characterisation of the selected halogen-free flame retardants. Literature data on acute toxicity and ecotoxicity tests of the selected halogen-free flame retardants was collected. The ecotoxicity data showed that a lack of data or contradictory data existed for halogen-free flame retardants that made it difficult to assess the alternative flame retardants and points to the need for reliable experimental data. This was further confirmed for data on specific end points based on a molecular and cellular level, with emphasis on geno-, endocrine-, and neuro-toxicity. Some of these toxicity end points were studied in ENFIRO to fill the data gaps on Ah-receptor, mutagenicity, thyroid hormone binding, endocrine disruption, and neurotoxicity.

Available data for physical-chemical properties for the selected non-halogenated flame retardants was reviewed. The physical-chemical data was used to assess environmental fate and behavior. It was found that estimation tools for organic substances exist but no reliable estimation tools were at the time available for inorganic substances, which means that the assessment of environmental occurrence of inorganic flame retardants was a major challenge. A review of the physical-chemical properties and the (eco)toxicity data for the alternative flame retardants was published (Waaijers et al. 2013).

Information on the economic aspects of the prioritized halogen-free flame retardants was collected as well. The focus was to give an overview about the flame retardants market and about the related industry which is highly influenced by recent trends. Pinpointed were those economic data that were collected through the ENFIRO project with the help of the project partners and the ENFIRO Stakeholder Forum members, in order to complete the Life Cycle Costing.

Table 4 List of selected commercial viable alternative flame retardants in combination with polymers that were studied in ENFIRO. List includes feedback from the ENFIRO Stakeholder Forum

Polymer materials	Mainly used brominated flame retardant	Applications	Halogen-free flame retardants selected
Epoxy resins	TBBPA	Printed circuit boards, Electronic components encapsulations, Technical laminates	Dihydrooxaphosphosphaphenantrene oxide (DOPO), Aluminium hydroxide (ATH), Fyrol PMP
Epoxy encapsulates	DecaBDE	Electrical encapsulating and casting	Melamine polyphosphate (MPP), Boehmite, Aluminium diethylphosphinate (Alpi), ATH, Zinc hydroxyl stannate (ZHS), Zinc stannate (ZS), Zinc borate (ZB)
HIPS/PPE	DecaBDE/ATO	Housings for business machines, dashboards, toys, equipments for refrigerator, telephones, and other consumer electronics	Resorcinol bis (biphenyl phosphate) (RDP), Bis phenol A bis (biphenyl phosphate) (BDP), Triphenylphosphate (TPP)
PC/ABS	DecaBDE/ATO	Housings for business machines, dashboards, toys, equipments for refrigerator, telephones, and other consumer electronics	RDP, BDP, TPP
Polyamide 6 Polyamide 6,6	Brominated polystyrene (BPS)/ATO	Electrical and electronic equipment, connectors, switches etc.; encapsulated electronic components	Alpi, MPP, ZB, ZS, Melamine cyanuarate (MC)

(continued)

Table 4 (continued)

Polymer materials	Mainly used brominated flame retardant	Applications	Halogen-free flame retardants selected
Polybutylene therephthalate (PBT)	Brominated polystyrenes/ATO	Electrical and electronic equipment, connectors, switches etc.; encapsulated electronic components	Alpi, Nanoclay (organo-clays based on montmorillonite, nano-MMT)
Ethylene vinyl acetate (EVA)	DecaBDE/ATO	Wire and cable	ATH, Magnisium hydroxide (MgOH), ATH coated with Zinc hydroxy stannate (ZHS), Boehmite
Textile coatings	DecaBDE/ATO	Protective clothing, carpets, curtains, upholstered fabrics, tents, interior in public transportation	Ammonium polyphosphate (APP), Pentaerythritol (PER), MPP, ZB
Intumescent coating: High impact polystyrene (HIPS)	DecaBDE/ATO	Housings of electronic products	Novel application to attempt to reach V(0) for pure HIPS with intumescent coating based on APP, PER, MPP

A schematic presentation of the most viable flame retardants was made per technical area, application, polymer, brominated flame retardants and alternative flame retardants, but a ranking was not possible as too many data gaps existed.

3.2.3 Hazard Exposure

Of the 14 alternative flame retardants that had initially been selected, seven were found to be less toxic than some of the brominated flame retardants and also that they accumulated less in the food chain. The environmental fate models that were used predicted that the organic halogen-free flame retardants would be found primarily in soils, sediments and dust and to a lesser extent in water and air. Controlled air emission experiments showed that all organic halogen-free flame retardants emitted from polymers at elevated temperature but not at lower temperatures. Leaching experiments showed that both halogen-free flame retardants and brominated flame retardants can leach to water. For some polymers no differences in leaching behavior were found between brominated flame retardants and halogen-free flame retardants, but some halogen-free flame retardant systems had higher leaching properties than polymeric based brominated flame retardants. The type of polymer is the main parameter determining the leaching behavior. Analysis of organic halogen-free flame retardants in dust from micro-environments and environmental samples showed highest concentrations on and around electronic equipment, in sediment and in sewage sludge.

3.2.4 Fire and Application Performance

All the selected alternative flame retardants fulfilled the regulatory fire test. A method was developed using intrinsic flammability properties as well as a simple method for characterizing the fire performance and fire toxicity of polymers using three parameters (fire spread, smoke/carbon monoxide, inefficiency of combustion). With this model a comparative fire performance assessment of halogen-free flame retardants versus brominated flame retardants could be made. An important finding was that halogen-free systems demonstrated clear benefits, for example less visible smoke, less toxic components in smoke and in some cases lower peak heat release rate. Regarding mechanical properties, the polymers with brominated and halogen-free flame retardant showed similar loss compared to the polymer alone. All formulations (halogen-free flame retardant and brominated flame retardant) showed equal or better performance regarding processability for injection moulding. For all polymer systems investigated a halogen-free flame retardant option was found. The results for the printed circuit boards showed that the halogen-free flame retardants were as good as or better compared to the reference printed circuit boards produced using brominated flame retardants. A novel intumescent coating system was developed for pure high impact polystyrene (HIPS), showing good fire performance results and excellent results were obtained for the industry fire standards relevant to the electronics industry as well.

3.2.5 Risk Assessment

The environmental and human risk assessments carried out during the ENFIRO project showed that the predicted environmental and human exposure concentrations were below the toxicity thresholds for the selected halogen-free flame retardants. However, the lower risk of halogen-free flame retardants compared to brominated flame retardants is mainly due to the lower hazards of the halogen-free flame retardants, and not due to a lower exposure. Reducing the leaching of halogen-free flame retardants from polymer materials is a next challenge for the development of new flame retardants.

3.2.6 Impact Assessment

The comparative life cycle assessment (LCA) of brominated flame retardant vs halogen-free flame retardants, using a laptop as case study showed that the waste phase was the most relevant phase. In particular, improper electronics waste treatment, leading to the formation of brominated dioxins from brominated flame retardants, had a strong negative impact on the LCA-scores. Overall the LCA performance of the halogen-free flame retardant scenario was better than for the brominated flame retardant scenario. The same life cycles were also evaluated on social criteria using a Social Life Cycle Assessment (S-LCA). Several hotspots are found in the raw material mining phase. In conclusion, ENFIRO showed that viable alternative flame retardants are available. Some halogen-free flame retardants showed less risk for the environment and human health, and show similar fire performance and technical application capabilities as brominated flame retardant

3.2.7 Conclusions from the ENFIRO Project

During the ENFIRO project, a unique approach to assess the data at three different levels was developed: the chemical (flame retardant), material and the product (Fig. 6).

The project followed a tiered approach (Fig. 7), starting in the first tier with a prioritization and selection of alternative flame retardants taking into account the viability of flame retardant production, application, flammability in product system,

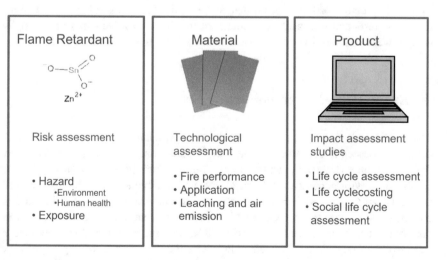

Fig. 6 ENFIRO's three levels of comparative assessment. Modified from Leonards (2011)

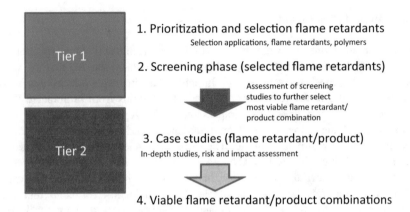

Fig. 7 ENFIRO tiered approach of screening and case studies. Modified from Leonards (2011)

hazards, and exposure of the flame retardants. This generated a list of viable alternatives and identified knowledge gaps. To fill some of the data gaps, screening studies of the selected flame retardants were performed. The screening studies focused on relatively rapid hazard characterization tests, exposure assessment modeling and fire performance tests.

Based on the evaluation of the screening results and literature information a further selection of viable flame retardants was narrowed down to be able to carry out in-depth studies on a selection (Tier 2). These studies covered chronic toxicity tests, neurotoxicity, a battery of in vitro tests, persistency, and monitoring of the behavior of the alternative flame retardants in the outdoor and indoor environments. In parallel, elaborated fire performance (realistic fire smoldering and flaming incidents) tests and technical assessments of the flame retardants in various applications were compared with traditional brominated flame retardant systems.

4 Evaluation of Interventions

The conclusions from the two case studies are presented below. In the ENFIRO case, the foreseen intervention has been to substitute one chemical (the brominated flame retardant decaBDE) with another (ammonium polyphosphate (APP) in combination with pentaerythritol (PER), melamine polyphosphate (MPP) and zinc borate (ZB)). In the SUPFES case, the one-to-one substitution of a hazardous chemical with a less toxic alternative has been one identified route. However, the possibility that the perceived requirements from customers turn out to be negotiable has also been considered, in a scenario where the customer is willing to make a compromise between environmental impact and technical functionality.

4.1   SUPFES Evaluation

In the case of DWR treatment of textiles the textile sector has since the beginning of the SUPFES project moved towards alternatives. The importance of SUPFES is therefore high and the results are being communicated along the study rather than after finalizing the project. The characterization of original substances in use was used to associate different types of PFAS with different chemical products. Figure 8 shows that the commercial chemical products labeled as "C6" chemistry also to a large extent contained "C8" chemistry and vice versa, as well as other fluorinated chain lengths.

The project has so far been able to conclude that when oil repellence is needed there are no viable alternatives to traditional long chain PFAS. At the same time the hazard assessment have shown both for long chain and to some extent short chain PFAS that they are rather alarming, especially due to their persistence. Regarding emission studies they do not under normal condition give rise to leakage during use. However, emission of stressed materials such as aging in UV and humidity may contribute to emissions.

Alternatives assessed do not offer the same level of technical performance. In addition some of them do emit during use and therefore lose their performance rapidly. They may also have to be added in higher amounts/concentration during application process. But indications are in most cases that non-fluorinated alternatives all in all give less environmental impact or health impact than fluorinated alternatives in general. Figure 9 shows the results from a GreenScreen (Clean Production Action 2014) chemical hazard assessment (Holmquist et al. 2016).

However, many of the alternatives, including short chain perfluorinated compounds, have not been thoroughly assessed to the same extent as long chain

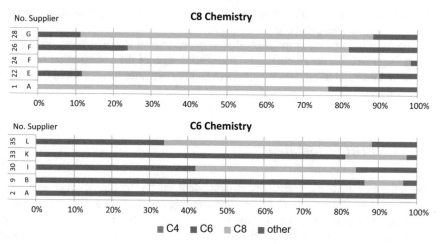

Fig. 8 Characterization of the commercial C6 and C8 products and association of different types of PFAS with different chemical products. Figure based on Van der Veen et al. (2016)

Substance	Hazard classification per endpoint											
	Human health								Ecotox		Fate	
	C	M	R	D	E	AT	ST	N	AA	CA	P	B
Benchmark												
PFOA	H	L	*H*	H	PEA	M	H	DG	L	*L*	vH	H
PFAS												
PFHxA	*L*	*L*	M	*M*	PEA	L	*M*	DG	L	*H*	vH	*L*
PFBS	DG	L	L	L	PEA	L	*L*	DG	L	L	vH	L
Silicones												
Short-chain silanols	DG	DG	DG	DG	DG	DG	DG	DG	DG	DG	DG	DG
DMSD	DG	DG	DG	DG	DG	DG	*M*	DG	DG	DG	*vH*	L
TMS	DG	*L*	DG	DG	DG	*M*	*M*	DG	L	DG	DG	L
D4	*L*	*L*	*L*	*L*	DG	*L*	*vH*	DG	L	*vH*	vH	*vH*
D5	*L*	*L*	*L*	*L*	DG	*H*	*H*	DG	L	L	vH	*vH*
Hydrocarbons												
Paraffin Wax	L	L	*L*	*L*	DG	L	DG	DG	L	L	vL	L
Nanotechnologies												
Dendrimers	DG	DG	DG	DG	DG	DG	DG	DG	DG	DG	DG	DG
Inorg. nanoparticles	DG	DG	DG	DG	DG	DG	DG	DG	DG	DG	DG	DG

Fig. 9 Hazard assessment for selected water repellent agent related substances that reach the environment via diffuse emissions, figure modified from Holmquist et al. (2016). Hazard classification abbreviations: vL = very low, L = low, M = moderate, H = high, vH = very high, PEA = potentially endocrine active, DG = data gap. Classifications in italics are of low confidence and in bold of high confidence. Classifications based on estimated data are marked with an asterisk (*). The endpoints are in order: Carcinogenicity (C), Mutagenicity and Genotoxicity (M), Reproductive Toxicity (R), Developmental Toxicity (incl. Developmental Neurotoxicity) (D), Endocrine Activity (E), Acute Mammalian Toxicity (AT), Systemic Toxicity and Organ Effects (incl. Immunotoxicity) (ST), Neurotoxicity (N), Acute Aquatic Toxicity (AA), Chronic Aquatic Toxicity (CA), Persistence (P) and Bioaccumulation (B)

perfluorinated compounds. Therefore, a few alternatives have shown hazardous properties in the same level as PFAS. For example, in the case of silicon-based water and dirt-repellent agents, such agents have recently caused concern due to the fact that both precursors and breakdown products have documented toxic, persistent and bioaccumulative properties.

The materials chosen in the study do not generate fiber loss which can otherwise be an important source of emission. Thus, ecodesign of the combination of alternative and material as well as intended use and performance levels should therefore be considered.

4.2 ENFIRO Evaluation

The efficiency of flame retardants is dependent on the textile polymer systems. Therefore, two different fiber types were investigated with two different coating polymers frequently used on the market. Polyamide (PA) weave and polyether terephtalate (PET) weave (also referred to as polyester) are used as filament plain weaves. The coating polymers were water based emulsion systems without cross linkers. The polymers in the two emulsions are acrylic respectively polyurethane

(PUR). A reference was also made for comparing the studied systems with best practice. This best practice was composed with decaBDE/antimony trioxide system. Dispersions with alternative flame retardants (ammonium polyphosphate (APP), melamine polyphosphate (MPP), pentaerythritol (PER)), coating of substrates, and fire testing of the coatings were performed. Test vehicles were tested for fire retardant behavior, peel adhesion strength between weave and coating, tensile properties of pure coating and friction. Representative textiles were used in this study and test vehicles were fire tested according to appropriate fire standards, required for the specific use. Dispersions of acrylic and polyurethanes were used for coatings on PET and PA weave.

Results showed that for suitable flame retardancy for PUR on PET weave 30% of MPP is needed. The combination with APP and PER is not more effective. A formulation of three halogen-free flame retardants (MPP, APP, PER) gives improved extinguishing compared to decaBDE. This halogen-free flame retardant combination is suitable for PUR on PA weave. The minimum amount of halogen-free flame retardant needed is 20% in solid coating, and the effectiveness is similar to decaBDE. Acrylics on PET weave can be flame proofed with 30% APP, but the combination with MPP and PER is not more effective. In this case also the combination of MPP, APP and PER gives similar extinguishing compared to decaBDE systems. For acrylics on PA weave none of the tested halogen-free flame retardants seem to be effective.

Tensile tests were performed on the coating according to modified SS-EN ISO 13934-1:1999. Bromine containing formulations show high tensile strength and maintained or improved elongation at break for both PUR and acrylic. The halogen-free flame retardant formulations are good for acrylic but poor elongation at break for PUR. The test also showed that the bromine formulations make a more flexible coating which is an advantage in many cases. The peel tests showed that optimized halogen-free flame retardant formulations (MPP, APP, PER) with PUR on PET weave had a 57% drop compared to decaBDE systems. The PUR on PA weave system performed better with the halogen-free flame retardants and showed only a 17% drop of peeling compared to the decaBDE. Interestingly, the acrylic on PET and PA weave textile system gave no coating with the decaBDE system but the MPP, APP, PER system gave adhesion.

Based on literature information, databases, and the ENFIRO hazard assessment, seven of the selected halogen-free flame retardants showed to have less issues of toxicity concern (listed as "no immediate concern" in Table 5) than some brominated flame retardants. For two of the halogen-free flame retardants, the results varied between aquatic toxicity studies in the literature (moderate-low and high-low toxicity, respectively). This variation may be due to the amount of triphenyl phosphate (TPP) present in the technical products; TPP is a by-product and know to be very toxic for aquatic organisms. The indication of high toxicity impact does cause concern for environment and humans for these two substances.

Also the results for TPP by itself cause concern: TPP is classified according to REACH as very toxic to aquatic organisms, may cause long-term adverse effects in

Table 5 Summary information on the hazard characterization of the selected halogen-free flame retardants

Risk category	Substances	Comment
No immediate concern	Ammonium polyphosphate (APP) Aluminium diethylphosphinate (Alpi) Aluminium hydroxide (ATH) Melamine polyphosphate (MPP) Dihydrooxaphosphaphenanthrene (DOPO) Zinc stannate (ZS) Zinc hydroxystannate (ZHS)	• Inorganic and organic substances with low acute (eco)toxicity and no bioaccumulation potential • Chemical stability required for application results in limited degradation (persistence) • Stannates: the (neuro-)tox effects found with in vitro cell based systems were not confirmed with animal studies (in vivo, mice studies), probably due to low bioavailability, therefore no immediate concern
Some concern for environment and humans	Resorcinol bisphosphate (RDP) Bisphenol-A bisphosphate (BDP)	• RDP toxicity to aquatic organisms is main concern, may be linked to by-products (TPP). Low and high toxicity are found for same test species, which is may be due to batch differences in the amount of TPP present as by-product • BDP is persistent
Of concern, risk assessment necessary	Triphenyl phosphate (TPP) Nanoclay (Cloisite)	• TPP very toxic to aquatic organisms is main concern, potential endocrine effects • Nanoclay showed strong in vitro neurotoxicity. May be due to the nanoparticle coating, additional studies needed. Information on the leaching behaviour of nanoclays from polymers is also needed

the aquatic environment, and needs to avoid release to the environment. Further, bisphenol-A bisphosphate (BDP) is a persistent compound. Finally, a third compound that is of concern and needs further study is the nanoclay (nano-MMT), which showed a strong in vitro neurotoxicity effect. Also the fate (leaching) of this compound from polymers needs further study. For an overview of the hazards see Table 5.

5 Conclusions

This chapter has presented a model for practical substitution where in addition to evaluating the environmental and health performance of alternatives, the technical and economical performance of alternative chemistries are also included.

This chapter also provided two examples of practical substitution using the model, referring to the EU FP7 program ENFIRO and the FORMAS funded SUPFES. The main conclusion is to work across different disciplines to generate knowledge about technical performance in parallel to a thorough evaluation of the environmental performance and health aspects. The technical performance evaluation should also include a discussion about acceptable performance level and different relevant or newly developed performance criteria.

In general, a substitution model including evaluation for both technical as well as environmental and health performance requires an interdisciplinary approach to create and/or suggest feasible alternative solutions. From a sustainability perspective it is vital that the substitution has a net positive effect when considering the entire life cycle. Thus, all relevant impact categories thus have to be considered to avoid new solutions that have less impact from a toxicity perspective but, for instance, have detrimental climate impact. But it is also important to avoid substituting an old and well-known toxic chemistry with a new, less known chemistry that may still be equally toxic.

References

AAFA (2015) AAFA restricted substance list (WWW Document). https://www.wewear.org/rsl/. Accessed 26 Oct 2015

AFIRM (2015) AFIRM supplier toolkit

Andersson H, Harder R, Peters G, Cousins I (2014) SUPFES: environmental risk assessment on short-chain per- and polyfluoroalkyl substances applied to land in municipal sewage sludge. In: 6th international workshop on per- and polyfluorinated alkyl substances—PFASs. Idstein, Germany

Backe WJ, Day TC, Field JA (2013) Zwitterionic, cationic, and anionic fluorinated chemicals in aqueous film forming foam formulations and groundwater from U.S. Military bases by nonaqueous large-volume injection HPLC-MS/MS. Environ Sci Technol 47(10):5226–5234

Baumann H, Tillman A-M (2004) The Hitchhiker's guide to LCA. Studentlitteratur, Lund, Sweden

Bergman Å, Heindel J, Jobling S, Kidd K, Zoeller RT (2013) State of the science of endocrine disrupting chemicals, 2012. Toxicol Lett. United Nations environment programme and the World Health Organization, New York and Geneva. doi:10.1016/j.toxlet.2012.03.020

BLUESIGN® (2017) BLUESIGN® (WWW Document). http://www.bluesign.com/. Accessed 24 Aug 2013

ChemSec (2017) SIN list (WWW Document). URL Canadian Hazardous Products Act (Phthalates regulation SOR/ 2010-298). Accessed 20 Mar 2016

China Ministry of Environmental Protection (MEP) (2010) Provisions on the environmental administration of new chemical substances in China. Order No. 7 of the Ministry of Environmental Protection (MEP). China

Clean Production Action (2014) GreenScreen® hazard criteria (WWW Document). http://www. greenscreenchemicals.org/info/guidance-and-method-documents-downloads:GreenScreen®. Accessed 31 Jan 2017

Dubois A, Gadde L-E (2002) Systematic combining: an abductive approach to case research. J Bus Res 55:553–560

Ecolabel Index (2016) Ecolabel index (WWW Document). http://www.ecolabelindex.com/ ecolabels/?st=category,textiles. Accessed 24 Aug 2013

European Chemicals Bureau (2007) Review on production processes of decabromodiphenyl ether (decaBDE) used in polymeric applications in electrical and electronic equipment, and assessment of the availability of potential alternatives to decaBDE. Brussels, Belgium

Commission European (2008) Regulation (EC) No 1272/2008 of the European parliament and of the council of 16 December 2008 on classification, labelling and packaging of substances and mixtures, amending and repealing Directives 67/548/EEC and 1999/45/EC, and amending regulation (EC). Off J Eur, Union 353

Commission European (2006) Regulation (EC) No 1907/2006 of the European parliament and the council of 18 December 2006 concerning the Registration, Evaluation, Authorisation and Restriction of Chemicals (REACH), establishing a European chemicals agency, amending directive 1999/45/E. Off J Eur Union L396:0001–0851

Commission European (2011) Directive 2011/65/EU of the European parliament and of the council of 8 June 2011 on the restriction of the use of certain hazardous substances in electrical and electronic equipment. Off J Eur Union L174:88–110

GOTS, 2017. Global Organic Textile Standard (GOTS)

Government of India (2012) Draft national chemicals policy (Draft NCP-2012)

Holmquist H, Schellenberger S, van der Veen I, Peters GM, Leonards PEG, Cousins I (2016) Properties, performance and associated hazards of state-of-the-art durable water repellent (DWR) chemistry for textile finishing. Environ Int 91:251–264

Herzke D, Olsson E, Posner S (2012) Perfluoroalkyl and polyfluoroalkyl substances (PFASs) in consumer products in Norway—A pilot study. Chemosphere 88(8):980–987. https://doi.org/10.1016/j.chemosphere.2012.03.035

Howard P, Muir DCG (2010) Identifying new persistent and bioaccumulative organics among chemicals in commerce. Environ Sci Technol 2277

Knepper TP, Frömel T, Gremmel C, van Driezum I, Weil H, Vestergren R, Cousins IT (2014) Understanding the exposure pathways of per- and polyfluoralkyl substances (PFASs) via use of PFASs-containing products—risk estimation for man and environment. Texte 47/2014. Dessau-Roßlau, Germany

Leonards P (2011) Life cycle assessment of environment-compatible flame retardants (Prototypical Case Study), ENFIRO. NORMAN Network of Reference Laboratories for Monitoring Emerging Environmental Pollutants New Bulletin 1–20

Munn K (2011) The chemicals in products project: case study of the textiles sector. Switzerland, Geneva

Norwegian Pollution Control Authority (SFT) (2009) Guidance on alternative flame retardants to the use of commercial pentabromodiphenylether. Oslo, Norway

OECD (2013). OECD/UNEP Global PFC Group. Synthesis paper on per- and polyfluoronated chemicals (PFCs)

OECD (2015) Substitution and alternatives assessment toolbox (SAAT) (WWW Document). www.oecdsaatoolbox.org/. Accessed 26 Oct 2015)

OEKO-TEX® Association (2017) OEKO-TEX® standard 100 (WWW Document). https://www.oeko-tex.com. Accessed 24 Aug 2013)

Olsson E, Posner S, Roos S, Wilson K (2009) Kartläggning av kemikalieanvändning i kläder. Mölndal, Sweden

Outdoor Industry Association (2014) Chemicals Management Module (CMM) (WWW Document). https://outdoorindustry.org/advocacy/corporate-responsibility/chemicals-management-module/. Accessed 1 June 2016

Posner S (2009) ENFIRO. Life cycle assessment of environment-compatible flame retardants. WP2 prioritization and selection

Posner S, Roos S, Brunn Poulsen P, Jörundsdottir H, Gunnlaugsdottir H, Trier X, Astrup Jensen A, Katsogiannis A, Herzke D, Bonefeld-Jörgensen E, Jönsson C, Pedersen G, Ghisari M, Jensen S (2013) Per- and polyfluorinated substances in the Nordic Countries. TemaNord, Nordic Council of Ministers. doi:10.6027/TN2013-542

Quinete N, Orata F, Maes A, Gehron M, Bauer K, Moreira I, Wilken RD (2010) Degradation studies of new substitutes for perfluorinated surfactants. Arch Environ Contam Toxicol 59:20–30

Roos S (2016) Advancing life cycle assessment of textile products to include textile chemicals. Chalmers University of Technology, Inventory data and toxicity impact assessment

Roos S (2015) Towards sustainable use of chemicals in the textile industry: how life cycle assessment can contribute. Chalmers University of Technology, Gothenburg

Roos S, Jönsson C, Posner S (2017) Nordic textile initiative—report on labelling of chemicals in textiles. Denmark, Copenhagen

SAC (2017) Sustainable Apparel Coalition (SAC) (WWW Document). http://apparelcoalition.org/the-higg-index/. Accessed 24 Aug 2013

South Korean Ministry of the Environment (2011) Act on the registration and evaluation of chemicals (K-REACH). South Korea

Swedish Chemicals Agency (2014) Chemicals in textiles—risks to human health and the environment. KemI Report 6/14. Stockholm, Sweden

Swedish Chemicals Agency (2007) The substitution principle. Stockholm, Sweden

Swedish Chemicals Agency (2006) Hexabromocyclododecane (HBCDD) and tetrabromobisphenol-A (TBBPA). Government Commission Report, Stockholm, Sweden

Swedish Chemicals Agency (2004a) KemI Rapport 6/04—Information om varors innehåll av farliga kemiska ämnen. Sundbyberg, Sweden

Swedish Chemicals Agency (2004b) Survey and technical assessment of alternatives to decabromodiphenyl ether (decaBDE) in textile applications. Stockholm, Sweden

Swedish Chemicals Group/Swedish Textile Importer's Association (2016) Chemicals guidance. Mölndal, Sweden

UNECE (2006) Overview of existing information on PFOS production, use, emissions and pathways to the environment and cost/benefits with alternatives/substitutes. Bern, Switzerland

UNEP (2006) Strategic approach to international chemicals management (SAICM). Switzerland, Geneva

UNEP, POPS, POPRC.8, INF, 17 (2012) Technical paper on the identification and assessment of alternatives to the use of perfluorooctane sulfonic acid in open applications. Switzerland, Geneva

United Nations (2002) Report of the world summit on sustainable development, Reissued edn. United Nations publications, New York

van Leeuwen SPJ, van Velzen MJM, Swart CP, van der Veenm I, Traag WA, de Boer J (2009) Halogenated contaminants in farmed salmon, trout, tilapia, pan- gasius, and shrimp. Environ Sci Technol 43:4009–4015

Van Der Veen I, Weiss JM, Hanning AC, De Boer J, Leonards PEG (2016) Development and validation of a method for the quantification of extractable perfluoroalkyl acids (PFAAs) and perfluorooctane sulfonamide (FOSA) in textiles. Talanta 147:8–15. doi:10.1016/j.talanta.2015.09.021

Waaijers SL, Kong D, Hendriks HS, de Wit CA, Cousins IT, Westerink RHS, Leonards PEG, Kraak MHS, Admiraal W, de Voogt P, Parsons JR (2013) Persistence, bioaccumulation, and toxicity of halogen-free flame retardants. Rev Environ Contam Toxicol 222:1–71

Wang D-G, Norwood W, Alaee M, Byer JD, Brimble S (2013a) Review of recent advances in research on the toxicity, detection, occurrence and fate of cyclic volatile methyl siloxanes in the environment. Chemosphere 93:711–725

Wang Z, Cousins IT, Scheringer M, Hungerbühler K (2013b) Fluorinated alternatives to long-chain perfluoroalkyl carboxylic acids (PFCAs), perfluroalkane sulfonic acids (PFSAs) and their potential precursors. Environ Int 60:242–248

Weiss JM, Van der Veen I, De Boer J, Van Leeuwen SPJ, Cofino W, Crum S (2013) Analytical improvements shown over four interlaboratory studies of perfluoroalkyl substances in environmental and food samples. TrAC—Trends in Analytical Chemistry 43:204–216. https://doi.org/10.1016/j.trac.2012.10.005

ZDHC (2014) Roadmap to Zero Discharge of Hazardous Chemicals (ZDHC) (WWW Document). http://www.roadmaptozero.com/. Accessed 1 June 2014

Sustainable Wet Processing—An Alternative Source for Detoxifying Supply Chain in Textiles

P. Senthil Kumar and E. Gunasundari

Abstract This chapter discusses the sustainable wet processing techniques and their environmental impacts in the textile industries. Wet processing is a main sector in textile industries, which affects the end product and their quality of textiles. Large amount of water, chemicals and energy are required for various stages of wet processing operation. In this wet processing, Water is used as the solvent for the chemicals and dyes, because of its low price and availability. But, during the process, water gets polluted with chemicals and unspent dye stuffs and gives an end product as effluent. The toxic effluent is not easy to treat or biodegrade and is harmful to humans and animals. This kind of contamination and health problems arises normally in the conventional method of wet processing. So, the alternative methods are necessary to improve the sustainability of the textile wet processing. In the recent time, the new eco-friendly methods have been developed and are preferred mostly instead of conventional methods. Plasma, ultrasonic, laser, biotechnology digital inkjet printing are the new innovated eco-friendly technologies, which provide more advantages to wet processing. In these methods, there are no any harmful chemical, wastewater and mechanical hazards to textiles, etc. This study also clearly discusses the various stages of wet processing operations such as desizing, scouring, bleaching, mercerizing, dyeing, finishing and printing with a new innovated trend and their eco-friendly procedures and technologies on the wet processing.

Keywords Wet processing · Chemicals · Energy and eco-friendly methods

P. Senthil Kumar (✉) · E. Gunasundari
Department of Chemical Engineering, SSN College of Engineering, Chennai 603110, India
e-mail: senthilchem8582@gmail.com

© Springer Nature Singapore Pte Ltd. 2018   37
S.S. Muthu (ed.), *Detox Fashion*, Textile Science and Clothing Technology,
DOI 10.1007/978-981-10-4876-0_2

1 Introduction

Textile industry is the main industrial sector all over the world and is one of the oldest industries. In past few decades, the profit in this sector has extremely high because of the fast fashion trend and increased textile consumption. According to United Nations Commodity Trade Statistics Database, in 2013, the worldwide market was around $772 billion in the textile and apparels exports. China, India, Italy, Pakistan, Bangladesh and Germany were the major exporters in 2013 (Economic Times 2014).

In the wet processing, natural or man-made (synthetic) fibers are used as the raw material in the form of yarn, fabric and garments. Substantial quantity of the energy, water, different types of chemicals and auxiliaries are used for the treatment of fabric. Later, the more amount of polluted effluent is discharged from the process, as it comprises of unfixed dyes and chemical auxiliaries that are dangerous to human health and ecosystem. So, the separate treatment requires for this effluent before the disposal. The discharge of wastewater is depends the production of textiles. For example, every year, over 2.5 billion tons of waste water discharges in the textile industry, when the 54% of the world textile production in China (Anon 2015).

To overcome these problems, the textile industries need to modify the process, technology in environment friendly manner. There, the eco-friendly textile has been produced by the using azo free dyeing, low impact dyeing and bio-processing technique. In 1998, Anastas and Warner was introduced the "green chemistry" concept. The sustainable production of fabric can be achieved by the "green chemistry" concept (Anastas and Warner 1998). The usage of eco-fibers or organic fibers in processing is the key factor to reduce the environmental impacts. Organic cotton corn, bamboo, hemp, soy bean, milk, pineapple fiber, tea, jute fiber, banana leaf fiber, organic wool, and organic silk are the most commonly available eco-friendly fiber. The sustainable chemicals need to replace the toxic chemical to avoid the health and environmental issues. The sustainability can be achieved by the optimizations of water, energy, chemical, chemical substitution, process modification and equipment modification, etc. therefore, a small volume waste only produced by this method. In this chapter, sustainable wet processing, innovations and technology development, eco-friendly dye and fibers has been explained. Various technologies such as plasma, ultrasonic waves, enzyme, digital printing, foam finishing, bio-based dyeing technologies involves in the eco-friendly production.

2 Environmental Impact of Wet Processing

There are several sustainable problems available in the textile wet processing operations. During textile processing, non-biodegradable and eco-unfriendly, extremely chemical-intensive goods are used and the discharged effluents from textile mill containing more amounts of unfixed dyes and the toxic chemicals. The advanced treatments are required for the separation of toxic chemicals from the wastewater however the complete removal of chemicals is not possible. In the recent years, the wet processing operations are performed in the developing or less developed countries. The wastewater released from the wet processing is directly discharged into the environment without any further treatment, which affects the humans and animals. The environment effects of wet processing are clearly explained in the following subdivisions.

2.1 Importance of Water in Textile Wet Processing

In the worldwide, water is the one of an essential natural resource for both humans and industries. They cannot survive without water and have no replacement for it. All around the world, twenty percentage of water has to be consumed by various industries. In the textile industries, a huge quantity of water is required for the textile wet processing operations. This water obtained from different sources like surface water from lakes and rivers and subterranean water from wells. In these operations, water, this is generally consumed as a medium and used as a solvent for the solutions of chemicals. A large quantity of water spent as a cleaning agent during washing and cleaning purpose. For the steam generation also, which is utilized to increase heat in the process bath. From mill to mill, the requirement of water will vary based on the weight of the textile material, equipment type, processing technique and type of dyes and finishing agents during the wet processing. At the end of each process, the more amount of water is consumed for the washing of goods (fabrics). For one kg of fabric material, 50–100 L of water is needed for processing. In 1980, the average water consumption in the textile mill was 150–200 m^3 per ton of finished goods stated by Beckmann and Pflug. 4.3% of the total wet processing costs spent for wastewater treatment and water supply in that same year. Sometimes, water is spent needlessly because of broken or missing valves; hoses left running, circulation of cooling water even though machines, water cooler, and toilets are not working probably (Beckmann and Pflug 1983).

In continuous preparatory processes, the effluents released from desize J-box and caustic washer can be recycled and reused easily through the continuous waste stream. The characteristics of wastewater in wet processing are moderately constant. The wash water discharged from the caustic washer having a caustic substance that influences the elimination of sizing chemicals. However, in the batch operation, the reuse and recycle of wastewater is not easy because of the storage facilities, non-continuous characteristics of wastewater and higher liquor ratios (Smith 1986). A large amount of water used in preparatory process and it has to be reduced by flow reduction and counter current flow. Before dyeing, the necessity to wash off impurities in the textile material using water is to get a good dyeing uniformity. In the dyeing operation, the quantity of water requirement varies with the types of dyeing machinery used. Various low liquor ratio-dyeing machines are manufactured for dyeing fabric to conserve water. In general, the definition for the liquor ratio is the ratio of the quantity of water (in pounds) in the exhaust dye bath to the quantity of fabric material. Compare to the dyeing bath, the large quantity of water is used for washing during dyeing operation. However, in washing, the water used is not affecting the liquor ratio. Therefore, we cannot say that the minimum quantity of water consumed for the dyeing bath. The effectiveness of water wash is enhanced with number of washes using a small quantity of water when compare to the one-time fabric wash with large volume of water. The extra water in the fabric required to eliminate before every wash to avoid contamination of wash water.

After wet processing operations, the wastewater is discharged into public sewage or on open land along with unfixed dyestuff and chemicals that generally is called it as effluent. The quantity and quality of the effluent differs based on the mixture of chemicals like organic and inorganic pollutants in textile industries. These pollutants present in the wastewater increase biological oxygen demand (BOD), chemical oxygen demand (COD), total dissolved salt (TDS), total suspended solid (TSS), pH and decrease the dissolved oxygen (DO) content. The strong color observed in wastewater because of its aesthetic character. These complex mixture of pollutants affecting the aquatic animal forms and as well as the formation bioaccumulation to create toxic environment. So, the quality of water is to be needed to recover using some basic techniques such as filtration, flocculation, clarification, aeration, membrane filtration, etc.

2.2 Energy Consumption

In the textile industry, energy is consumed as fuel (oil, coal, natural gas and LPG) for the boilers to generate steam, as electrical source for cooling and temperature controlled systems, machinery and lighting. During dyeing process, the maximum of the energy has been consumed for heating the dyeing bath. The cellulose fiber can be dyed using lower temperature. However, for polyester more temperature may need for dyeing operation. The temperature requirement for dyeing is based on the type of dyes, fibers and process condition. Different machines such as drying machines, thermosol processes and stentering machines in wet processing are also consuming more energy for running machines (O Ecotextiles 2009). Thermal energy consumption is more in wet processing compare to the electrical energy consumption, which is mainly for heating of water. The energy consumption increased with increasing the water consumption. The energy is conserved by the proper chemical methods, process and machine modification and novel technologies. In steam pipes, energy is wasted by the poor maintenance, improper insulation and leakage in steam pipes. Therefore, proper insulation is important to control the heat losses in the workplace.

2.3 Chemical Utilization

Large amount quantities of chemicals are utilized in various textile industries and are surfactants, lubricants, cleaning agents, defoamers and some specialty chemicals. These chemicals are hazardous to the environment as well as too human because of toxic, carcinogenic, mutagenic properties. So, the offending chemicals are required for wet processing instead these chemicals. For example, in the dyeing and printing process, formaldehyde containing dye fixing agents are used along with dispersive or reactive dyes, which give eye and skin irritation. So, these chemicals should be replaced with the non-formaldehyde based chemicals (Arputharaj et al. 2015). However, the majority of the textile industries are often using the carcinogenic and toxic chemicals even it has health and safety issue between consumer as well as worker who are working in the dyeing industries. Chemicals used for the textile wet processing also harmful when these are left it in wastewater. This wastewater effluent pollutes the land and water bodies. Therefore, it is essential to reduce through chemical substitution the use of chemicals that result in harmful wastes. The chemicals must be reused, recycled for processing when no useful chemical substitutes can be obtained for the toxic chemicals.

3 Textile Wet Processing

In worldwide, Cotton is the most widely utilized natural fiber by various textile industries and have nearly about 40% share of the total global fiber consumption of cotton. The textile material has been processed by using four major steps such as preparatory pretreatment, dyeing, printing and finishing. These are discussed clearly in the following subdivisions.

3.1 Preparatory Pretreatment

Before preparatory pretreatment, first, the yarn materials are going for the sizing process. In the sizing process, the sizing chemical is incorporated into yarn, which is acted as adhesive substance. Sizing agents such as starch, polyvinyl alcohol (PVA), polyvinyl acetate, carboxymethyl cellulose (CMC) and gums are used to improve the properties of the warp yarn, like smoothness, abrasion resistance, and the tensile strength (Jones 1973). Therefore, during the weaving process, it will provide an ability to withstand the mechanical forces.

Then, the singeing operation has to be carried out to remove hairy fiber, lose fiber and protruding fibers on the surface of the woven fabric. This has to be done because; small fiber balls began to form on the surface of the fabric after the several washing of the cloth. Three common techniques are available in the singeing process and classified as gas flame, roller and hot plate techniques. In the gas method, the woven fabric materials are allowed to pass through gas fired burners. In roller method, the heated rotary copper cylinders are used for singeing. Likely, for the hot plate method, the heated plates are used at the speed of 150–250 yards/min.

The untreated textile material, also called as "grey textiles", is treated using different treatment processes and chemicals in the pretreatment department. The pretreatment process includes desizing, scouring, bleaching, and mercerization. In the pretreatment process, Desizing is the first preparatory pretreatment process. It is the process to degrade or solubilize the size fabric material with the help of a desizing agent. The main reason for desizing is that the naturally sizing ingredient such as starch present in the textile, which hinders the impregnation of dyes into fabrics in dyeing and printing processes. Therefore, the removal or conversion of starch into simple water-soluble product is important that may take place either by oxidation or by hydrolysis. In wet processing, three different desizing methods are used for desizing, such as acid desizing, oxidative desizing, and enzyme desizing. For degradation of the sizing materials, hot caustic soda or detergent treatment is also used but this treatment is not that much efficient compare to those treatment methods. Even though the need of water is probably less for the process, starch hydrolysis products have raised the biological oxygen demand (BOD) of the desizing effluent. In general, about 50% of water pollution is caused by high BOD wastewater that was produced during desizing and become unusable (Correia et al. 1994).

Scouring is used to eliminate the natural and add impurities like gums, waxes, lubricants and dirt, etc., from textile fibers and is carried out at high temperature and an alkaline environment. This process can be used to convert these impurities into soluble compound. So, the scouring effluent has excessive COD, BOD, TDS and alkalinity. Non-cellulosic substances such as pectin, wax and proteins from cotton fiber are removed in this process. The conventional scouring of cotton fiber needs a strong alkaline medium with the temperature 120 °C. In scouring bath, reducing agents, wetting agents, sequestering agents and emulsifiers are also applied to increase the effectiveness of the process. For example, if calcium, magnesium and iron present in textile material, soap and detergents (reducing and sequestering agent) are added to precipitate out it and then removed. At high pH, the sequesternants will create strong calcium, magnesium and iron (2+) complexes. Then, the reducing agents are used to reduce iron (3+) to iron (2+). This scouring process may be carried out either continuous or discontinuous manner. Silk material is also scoured with soap/surfactants under mild condition to remove the gummy substance (sericin) (Jones 1973). Sericin substances are also removed by hydrolysis with help of protease enzymes. Greasy matters from wool can also eliminated by the addition of surfactants and emulsifiers. For synthetic fiber, mild scouring is sufficient to remove dust, lubricants, oil, etc., because it will not produce the inherent impurities.

The natural colored matters from the textile materials are removed by bleaching process. This process is also used to improve the whiteness level of fabric. Bleaching agents are essential for bleaching process to destroy chromophores, probably breaking one or more double bonds into conjugated system. These bleaching agents are classified into three types and are oxidative, reductive and enzymatic bleaching agents. Comparing all these type agents, oxidative bleaching agents are most commonly used for the process. In the alkali solution, this agent is decomposed to form active oxygen that partly or completely destroys the natural colored matter in fabric material. The oxidative bleaching agents like hydrogen peroxide, sodium hypochlorite and sodium chlorite are most widely used in the decoloration process (Rott and Minke 1999). After scouring, the fabric is allowed to pass through the sodium chlorite bath for impregnation and then the traces of bleach are removed by washing with antichlor ($NaHSO_3$). The sodium chlorite is very successful at pH 4. Compare to the chlorine bleaching; peroxide bleaching is preferred mostly based on the permanent whiteness and environmental point of view. But the peroxide bleaching is comparatively very costly. In general, chlorite is used in the beginning of bleaching. Then, at the time of washing, antichlor ($NaHSO_3$) is replaced by the hydrogen peroxide, which is used in lower concentration. So, after bleaching, the permanent whiteness and economy of bleaching can be ensured.

Finally, before dyeing and printing process, the mercerization process is carried out with a strong caustic alkaline solution for the cotton material to improve the dye absorption capacity, luster, strength/elongation, smoothness, hand, dimensional stability, etc. This process along with a strong alkaline solution under tension forms the permanent swelling and shrinks in the fiber. Then the goods are washed with water to eliminate alkali and then will get a permanent silk-like luster (Osman 2007). Based on the concern of cost and desire result, caustic soda is mostly used for mercerization.

3.2 Dyeing and Printing Process

After the pretreatment process, the fabrics are all set to go for the dyeing process. In general, dyeing is the process of applying colors to fabrics or yarn that can be carried out either for various textile fibers, yarn and fabrics or for a finished product like apparels and garments. Dye solution is prepared by the combination of dye molecules and aqueous medium (water) which can be used for dyeing in neutral, alkaline or acidic environment. The dye molecules are chemically bonded with the textile fibers molecules. There are several stages are there for colorization of the textile fiber using dye molecules and the steps are as follows:

- A migration of dye molecules within the dye bath near fiber.
- Diffusion of dye molecules in the dye bath.
- Absorption of dye molecules from dye bath on the external surface of fiber.
- Diffusion of dye molecule through the interior surface of fiber.
- Then finally dye molecule well bond with fiber and is so called as "fixation" (Cegrra et al. 1992).

There are two types of dyes are available such as dyestuffs and pigments. Dyestuffs are dissolved easily with water or made to dissolve using an additive named as auxiliary chemicals. The reasons for the addition of auxiliary chemicals into the dyeing bath are mainly to govern the movement of the dyes, prevent foaming, leveling of a dye, etc. pigments are not solubilizing with aqueous medium. Therefore, a binding agent is essential to bind the dye with fabric or yarn. In these two types of dyes, dyestuffs are most widely used for dyeing operation.

Dyestuffs are classified into two major classes such as natural and synthetic dyes. Natural dyes are extracted from animals and plants. The most of dyes are produced from the plants like berries, leaves, roots, woods, etc. synthetic fibers are prepared from synthetic resources including earth mineral and petroleum by-products. Synthetic dyes are most commonly used for the dyeing process when compare to the natural dyes. The first class of fiber reactive dyes generally reacts with functional groups of fiber which are acid, basic, direct, reactive and mordant dyes. The second class of dyes are required chemical reaction, which are vat and sulfur dyes and, finally the third class dyes are called as a special dyes such as disperse, solvent and natural dyes.

Direct dyes are directly colored the cellulose fiber with the addition of salts. The discharge of effluents contains unfixed dyes and salts that pollute the environment. These dyes are cheap, different range of bright colors and easy to apply. But, they are not environment friendly. One type of direct dye is azo dyes, which is derived from carcinogenic amines. Later, the last decades of twentieth century, Germany and other European countries were banned some of azo dyes. Sulphur dyeing is carried out for less expensive fibers and garments. It gives only one color, Sculpture Black. Compare to all other types, no heavy metals are added into the dyeing bath during sulphur dyeing. However, the toxic effluents released from the process due to the use of sodium sulphide. The effluent has high pH and unpleasant odor. High quality textiles are to be dyed by these dyes. Alkali and reducing agents are mixed to solubilize these dyes. They are the least environmentally impactful dyes i.e. only 5–20% of dyestuff, oxidizing agent, reducing agents remain in the discharge effluents. However, they are more costly. This reactive dyes also has advantages like multiple colors, bright color and easy to apply. The effluent released from the process also has salts, alkali and unfixed dyes. Therefore, the synthetic dyes are having more advantage but these types of dyeing are non-ecofriendly and produce lots of toxic waste effluent. Reactive and direct dyes are most commonly used for cotton fiber due to the formation of covalent bonding between fiber and dyestuff. However, for the hydrophobic nature of polyester, the same kind of dyes is not used instead of that the insoluble disperse dyes are dyed because of the property difference between the cotton and polyester. Cotton/polyester fibers are dyed using both reactive and disperse dyes. This dyeing process can be carried out in the batch or continuous manner. Beam, package, jet and jig dyeing can be done in batch mode. Pad-batch dyeing is one of the special methods, which is using reactive dye for dyeing of cellulose fibers. In the continuous dyeing, the long length fabrics are passed through the dyeing bath. Steam is applied for the fixation of dyes onto the fabrics. Simultaneously, the excess dyes are washed with the help of water (Shukla

2007). The polyester is processed using disperse dyes at elevated temperature in a batch mode and then, the machine is cooled after dyeing. For the cotton portion, reactive dyes are refilled in machine. Finally, the excess dyes are removed from the fabric by washing or dye-extraction.

In the printing process, the thick paste and viscous form of dye and pigments are applied to prevent the migration of the dye in the fiber. Water-based ink, plastisol ink and some other pigments are involved in printing using different methods like direct, resist and discharge printing. In the direct printing method, the color pattern is directly printed on the surface of textile fabric. In the resist printing method, initially, a resist paste is applied in the desired printing pattern. Then dye is applied onto the fabric. The resist paste applied parts of fabric are not colored. In the discharge printing, the dye is applied on the fabric after, that a discharge paste is printed on the fabric. The availability of reducing agent in the discharge paste influences the removal of dye from fabric. This type of printing is not eco-friendly nature due the presence of formaldehyde in the discharge paste. Washing is required for this printing to eliminate the by-products. Mostly a rotary screen, flat screen or engraved roll is used for the printing color onto the fabrics (Ramesh Babu et al. 2007). High temperature is required to fix the printing dyes on the fabrics and as well as for drying operation. After printing, the unfixed dyes and pigments are released as effluent using water and some organic solvent just like in dyeing. The presence of toxic chemicals in the wastewater is harmful to the environment, wildlife and humans. So the proper waste treatment is necessary before the discharge into the environment.

3.3   Finishing Process

There are two types of processes can be carried out in finishing and are mechanical and chemical methods. The most popular mechanical process is calendaring which is provide the pressure to the fabric between rollers and form a flat and smooth fabric surface. Sharp steel points used for abrading the fabric surface that is called it as 'napping'. Compressive shrinking process is also used to prevent shrinkage of wet fabric. In the chemical finishing processes, the characteristics of the textile materials are improved by the addition of finishing chemicals. The formaldehyde-based chemicals are widely used such as softeners, dye fixing agents and cross-linking agents. These chemical agents are given easy care and better durability to the textile goods. The disadvantage of the use of formaldehyde-based chemicals may cause the skin rashes, eye irritation. The International agency for Research on Cancer (IARC) reported that these chemicals are 'carcinogenic to humans'.

In recent year, polymeric finishes have been used for resisting the water, stain and oil. This finishes are having long perfluoroalkyl chains with more than eight fluorinated carbons. The long-chain PFAAs are formed by the degradation of the residual raw materials in the environment, which is harmful to humans and animals. The antimicrobial agents are added to treat the textile fabrics get better the health, disinfection and besides to stop odor formation, Tirclosan was widely used as an antimicrobial agent but nowadays which is ignored due to the poisonousness. Different types of flame retardants also used for the finishing operation. Even if brominated and antimony oxide-based flame retardant are very harmful to humans, they are used for finishing and e.g. Polybrominated diphenyl ethers (PBDE) (Saxena et al. 2017). Thus, the main reason for finishing process can modify the shade of colored fabric and their fastness properties.

4   Recent Sustainable Development and Innovations in the Wet Processing

Nowadays, the textile industries in the various developed countries are facing more challenges due to the globalization process. Recent years, the news technologies and innovations have been developed for solving these problems to produce the sustainable textile material with economical manner. The innovations and developments in textile processing can be possible to substitute in various areas of wet processing such as pretreatment, dyeing and finishing. The sustainable developments are substantial for economy, quality, energy conservation and environmental considerations in the textile wet processing. In textile industries, wet processing is a large and essential sector, which has different methods to affect the appearance and quality of goods. The following innovation production techniques help to improve the sustainability of textile industry.

4.1   Eco-friendly Fibers

Eco-friendly fibers are produced from organic and eco-friendly materials and are classified as follows: organic cotton corn, bamboo, hemp, soy bean, milk, pineapple fiber, tea, jute fiber, banana leaf fiber, organic wool, and organic silk, etc. bamboo can grow well as it uses no pesticides and has a unique anti-bacteria and bacteriostasis bio-agent called "bamboo Kun". During the fiber production, these agents are closely packed with bamboo cellulose. Even after the several washing, fiber has bamboo Kun, which clears that the bamboo fibers having antibacterial, green and biodegradable properties. Organic cotton is also grown without using pesticides, herbicides or insecticides. A new type of colored cotton is grown for the production of the fibers, which is in the colors of green and brown. The advantage of this fiber is not easily fading, which is environmental and skin friendly.

4.2 Enzymes

The impact on the environment as well as product quality has been greatly benefited through the significant use of enzymes in textile wet processing. It is the one of most effective alternative to the use of hazardous chemicals and dyes in wet processing in the each stage. It will create the toxic free environment and produce the sustainable textile material. In the enzymatic processes, moderate conditions and the lesser energy are sufficient that leads to the reduction of the green house gas emissions from the power stations. The water and chemical consumption also reduced by using enzymes used in the every wet processing operations. The byproduct formation also minimized by the enzymes that offers nontoxic and eco-friendly environment. More than five thousand known enzymes are widely available in the world, in which, only about 75 enzymes are mainly applied in the textile wet processing industries. Hydrolases and oxidoreductase are most important used in textile industries and are classified as cellulases, amylases, proteases, lipases and pectinases. Alpha amylase enzymes effectively reduce starches cellulosic fibers into glucose than the acid desizing in desizing or bio polishing in the pretreatment processing (Saxena et al. 2017). These enzymes are applied to minimize the fiber breakage, effluent load and the consumption of water, energy and chemicals. One drawback of these enzymes is that the increase of BOD content in effluent. Instead of alkaline scouring, alkaline pectinase enzyme-based process is used for cotton fabric under moderate temperature to enhance the uniform dye uptake, less damage and energy saving.

The combination of desizing and scouring operation can be carried out to control fiber loss. For example, the non-cellulosic impurities are separated by the combination of enzyme-based scouring and activator-assisted bleaching in the single bath under moderate temperature. In bleaching, laccase can be used to bleach the coloring substance from wood pulp fiber. Flavonoids are giving the color to the cotton fiber, which is decolorized by the laccase system, and this system creates the stone wash effect to denims. Catalases have been studied for the removal hydrogen peroxide from fabric following bleaching (Saxena et al. 2017). For the removal of hydrogen peroxide from fiber material, the large quantity of water is used for washing to avoid difficulties in dyeing. The catalase enzyme with lesser amount of water eliminates these residual hydrogen peroxides.

4.3 Bio-Based Dyeing

The conventional dyeing processes are hugely affecting the environment even if they can give colorful textiles. The dyes present in the textiles risky to the wearers, workers who are in the industries and environment due to the presence of toxic chemicals such as dioxins and toxic heavy metals and formaldehyde that are carcinogenic in nature. To overcome this problem, novel methods are developed for the synthesis of coloring substance. For this bioprocess, no toxic chemicals require for the biological preparation of dyes. Small quantity of raw material, water and energies are sufficient in this biological process. This dye is named as "natural dyes or "bio-dyes", which are cost-effective, eco-friendly and durable. Generally, natural dyes can be produced from different sources such as plants, animals and microbes.

In the olden days, the natural dyes are used widely which is derived from animals, plants and minerals. In 1895, the first synthetic dye discovered and then followed by various dyes produced. Then, the usage of synthetic dyes is increased vastly due to the good color fastness, suitability and different color choice. Synthetic dyes used in the conventional dyeing process are grouped as direct dye, reactive dye, vat dye and sulphur dye. Then the natural dyes are started to use in the dyeing process owing to eco-friendly environment and may offer health benefits to the wearer. Natural dyes having a combination of chemical mixtures, which will differ, based on the variety, maturity, agro climatic variations. Therefore, the same kind of shade with natural dye is hard to reproduce for every dyeing operation and the similar result is produced under different pH and mineral content. For good color fastness properties, the metals are used along with dye, which is not eco-friendly. Thus, the heavy metals in the dyed fibers are avoided to improve the eco-friendly nature.

Compare to plant and animal-based dyes, microbial-based dyes have some advantages like fast growing nature and commercially standardized. Enzyme or fungi acts as a biocatalyst, which is bioprocessed with precursor in the bioreactor to produce the synthetic dyes. The fungus Basidiomycota is one of the effective biocatalyst, which oxidize the colorless precursor to produce colored bio-dyes in bioreactor. The production of the biocatalyst from the fungal culture is quite cheap and simple (Parisi et al. 2015).

The process of dyeing protein fiber (wool) can be carried out with anionic dyes. First, the dye molecules adsorbed on the surface of the fiber. After the adsorption, the diffusion takes place between dye molecules and fiber. The dye molecules bind with colorless acids in the fiber. Then the colored ions in dyes are displacing the colorless acid. Finally, the colored anions are bonded with reactive group in fiber. This fiber also contains amino group and carboxylic group. Comparing wool, it has a low dye fixation due to the less number of $-NH_2-$, $-COOH-$ group presence. Both anionic and cationic dyes are used for dyeing process. Swelling of fiber is more in water owing to that reason, the dyes diffuses easily inside the fiber (Parisi et al. 2015). Polyester and polyamide fiber are processed in the bio-based dyeing. In these, polyamide has the same dyeing properties and structure of natural fibers.

The bio-based process has been effective technology that increases the selectivity of a reaction and eliminates the downstream processing. Thus, this process is reducing the energy and material waste. The advantages of the eco-friendly bio-catalysis are lesser process steps, mild process condition (pH and temperature). The bio-based dyeing process can reduce the production toxic effluents and reduces the greenhouse gases. Generally, the bio-dye is in the form of liquid. However, the conventional method is using the dry powders. This dyeing process can be involved either batch or continuous mode, which is depends on the several factors like size of dyeing bath, the type of material (yarn, fiber, fabric and garment), common type of fiber and final product quality requirement.

4.4 Plasma Technology

Plasma treatment is the interesting water free treatment technology for the wet processing due to its usage and environmental tolerability. In this treatment, less chemicals and energy are enough to produce finished fabrics in a pollution free environment. Plasma is the fourth form matter as well as other three forms of matters includes solid, liquid and gas which was proposed by Sir Williams Crooke in 1879. Plasma is the partially ionized gas comprises of ions and was first discovered by Irving Langmuir in 1929. The atoms, radicals and electrons in plasma can be generated by the addition of external energy, which changes plasma as electrical conductivity to electricity. These effects can modify the fabric surface, which is depending on the type current used, temperature, pressure and the gaseous matter used and type of textile fiber. Cold plasma is used for the fabric treatment under the room temperature even, which has higher electron energy. Another one technique, plasma polymerization that is the process to polymerize the polymeric material onto the fabrics however, it is under development. Electrical discharge techniques are commonly used for the plasma treatment textiles. In this process, the gas is to be ionized in a controlled manner under vacuum conditions. Initially, a vacuum vessel is pumped to reach a low to medium vacuum pressure in the ranges of 10-2 to 10-3 mbar using rotary and root blowers. Then the gas is passed into the vessels via valves and mass flow controllers. Air, oxygen, nitrous oxide, argon and tetrafluoromethane are few of the gases used in processing (Sarita 2016).

The major advantages of such plasma treatments in textile processing are as follows:

- The bulk material properties are not changed by the plasma treatment, only the surface of the treated material affected.
- Due to the entrapment of byproduct, the effluents are not discharged into the environment.
- No hot water and chemicals instead of that O_2/He plasma or Air/He plasma is used for desizing operation.
- The wastewater and their treatment cost reduced due to the waterless treatment.
- The damage of the heat sensitive materials is abridged since the plasma occurs at room temperature.
- All kinds of fabric materials can be processed in this plasma treatment.
- Dyestuff fixation and leveling of dye properties are enhanced by the plasma technology. i.e., The plasma can imparts the anti-felting properties to the wool and improve the dye ability of the natural and synthetic fabric materials.
- No drying process required owing to the non-aqueous plasma treatment.

4.5 Supercritical Fluid Technology

Supercritical fluid technology has been another waterless treatment technology for the wet processing. In this treatment, gases are used instead of water, which is possible of changed into supercritical fluid. High pressure and temperature are required for supercritical fluid to dissolve with dyes. The most commonly usable supercritical fluid is CO_2 owing to their low cost, low viscosities and high diffusion rates, which influence the dyes to diffusion into the fabric. Carbon dioxide gas is converted into the supercritical fluid at low pressure and temperature. It has properties of both liquid and gas. In the dyeing operation, first, the dyestuff is mixed with supercritical liquid with increased temperature and pressure and then the mixture is transferred and diffused into the fabric. CO_2 and excess dyes are easily recycled after the process. When compare to hydrocarbon, carbon dioxide is most widely used as a solvent for the fluorinated compounds due to its solubility.

In Supercritical dyeing processes, the fabric sample is covered around a perforated stainless tube that is placed inside the autoclave. At the bottom of the vessel, the dyes powder is placed and closed tightly. In which, the gaseous CO_2 is continuously purged and preheated. Then CO_2 is compressed with aid of optimum working temperature and pressure. For 1 h, the same pressure is maintained and after that, the pressure is reduced to recycle CO_2 and dyes. Finally, the dry textile sample is removed from the vessel (Sarita 2016). The main advantages of CO_2 used in this treatment are inert, easy to handle, recover and reuse, availability, low cost, no waste generation and as well as eco-friendly. For example, in the dyeing process, the supercritical carbon is used as the medium instead of water for the polyester and polyamide using pure disperse dyes. This process is the effluent-free dyeing process. The major advantages of this technique are short dye cycle, waterless dyeing, effective leveling and no drying step.

4.6 Digital Ink-Jet Printing

Digital ink-jet printing has been established to replace the conventional process in the textile industry. Apart from having few advantages like prototyping, small run printing, customization and experimentation, this printing also does not fail to be within the budget. This digital ink-jet printing has better properties such color fastness and pattern quality that can be used for all type of the fabric materials. The advantages this technique are lesser energy and water consumption, higher dye fixation rate, no cleaning and washing required for equipment's and no wasted inks and dyes. In digital ink-jet printing, ink is directly sprayed on the surface of textiles with the help of nozzles without contact, which generally named as the non-contact technology. This printing process can be used to make different color designs using design data from the computer file without the need of any screens or running of heavy-duty machinery. However, it is not possible to get same shade and color what you see on screen. In the inkjet printing, a printing head releases more number of tiny drops of different color ink on textile substrate. These drops are combined to produce the photo-quality image. The quantity of drops ejected from printing head is depends on the nozzle size, the actual print head principle. After printing, heat/steam is applied for curing the ink. In the digital printing process, different kind dye inks are used such as disperse dye, acid dyes, reactive dyes, pigment ink, latex ink based on each type of fabric includes cotton, polyester, nylon, silk, etc. (Ibrahim 2012).

Disperse ink is widely used as a water-soluble ink which produce bright color to the textile. It is only worked with polyester and polyester/natural fibers printing. Both direct method and transfer methods are carried out to print the fiber using this ink. In the direct method, disperse dye ink is directly applied to print on the textile. In the transfer method, initially, this ink is printed on a roll of paper. After drying that, the paper is pressed on the textile material with help of calendaring roller. Then heat applies to vaporize the ink and then permanently absorbed on the textile (polyester and polyester/natural fiber). Transfer method is better than the direct method.

Acid dye ink is apt natural fiber like wool, silk, albumen fiber and as well as for polyamide textile, nylon and nylon/elastane and is having good color fixation and saturation properties. Medium to low viscous inks is used for piezo electric print heads. In digital printing, Reactive dyes are chemically bonded with textile. Only, medium viscous inks are suit for piezo print heads. Both types of inks is directly printed on the textile. After the printing, the unfixed dyes and ink-receptive coating are washed off using water. High heat and water are essential for these dyes (Eccles 2016).

Pigment inks are used mainly for direct printing of textile rolls and garments. Heat pressing is required for binding of pigment on textile later printing. These inks are commonly used for polyester, viscose, and leather cotton (Eccles 2016).

Latex-capable natural and synthetic material can be printed by using latex ink in digital printer which gives bright color. It is odorless and solvent free ink (Eccles 2016).

4.7 Ultrasonic Waves

Ultrasonic technology has been interesting technique in the textile wet processing. It can able to utilize and modify the wet processing. This ultrasonic wave have been used all wet processing operations to reduce the process time, minimize the usage of auxiliaries and improve the quality of the textile material. This technique used in the process is more eco-friendly. The main advantages of the ultrasound are and lesser processing time. The frequencies of ultrasonic waves range from 20 to 500 kHz. These frequencies are inaudible to human ear. The ultrasonic waves can be used in different part of fabric processing such as desizing, scouring, bleaching, dyeing and as well as for washing. Ultrasonic desizing is the energy saving method as compare to the conventional sizing and desizing. It can degrade the starch but not degrade the fiber. It gives better whiteness and wet ability. The scouring of wool in the neutral and very light alkali bath reduces fiber damage. 20 kHz frequency is sufficient for the peroxide bleaching of cotton fiber. It has been found that bleaching rate increased to increase the degree of whiteness as compared to the conventional bleaching. During dyeing, cotton uses the direct dye; the polyamide and acetate fiber use the disperse dye; wool uses the acid dyes. Ultrasound is more useful to the water insoluble dye to hydrophobic fibers. The dyeing is depends on the frequency of ultrasound. 20 kHz ultrasound is enough to make cavitations (Sarita 2016; Ibrahim 2012). The advantages of this method are lower consumption of energy, chemicals and time as well as lesser processing cost.

4.8   Innovations in the Machinery

Many advanced machineries have developed for textile wet processing. Water, energy and chemical consumption for the wet processing has reduced by these machineries. In the batch processing of fabric, low liquor ratio processing has to be carried out with lesser amount of water using the advanced machineries. This process not only reduces the water consumption and also save energy and chemicals as well as reduce the amount of effluents. These machineries are equipped with microprocessor-based controllers that minimize the carbon dioxide emission and energy consumption. Likewise, chemical consumptions in wet processing have reduced by the installation of an automated chemical dispensing system with processing machineries. Compare the batch processing, continuous processing have been often used in the textile industry. Continuous bleaching and dyeing ranges (CBRs) are equipped with prewasher, comi steamer, dosing system with automatic control, effective washing unit and dryers. This type of processing reduces the water, energy, space and operational cost.

In the dyeing process, high efficiency padder, vaccum application of dues and various dye application systems have been used for minimize the environmental impact of the dyeing process. Waterless dyeing technology has also introduced in the dyeing process. In this type, instead of water, air is used that reduces the liquor ratio and as well as effluent production. The dyed textile is not to go for drying owing to non-aqueous treatment. Therefore, energy consumption for processing also reduced. Knits and Closed HTHP jiggers have been developed for the dyeing of PET fibres using ultralow liquor ratio of 1:2 (Saxena et al. 2017). The usage of insulated steam pipes and machineries as well as heat exchangers conserve energy and create the less hot environment to the worker and effluent.

4.9   Recycling and Reuse of Process Inputs

Recycle and reuse of process input in wet processing can minimize the cost of process and as well as the environmental impact. Treatment of textile effluents can recovered the process inputs and reduces the colors and COD content using advanced membrane processes such as microfiltration, ultrafiltration, Nano filtration, reverse osmosis, adsorption and ion exchange processes. A small number of chemicals from the waste effluents are only recovered and reused. Alkali containing mercerizing wash liquor can be used for scouring and bleaching operations. A large quantity of water can be used for wet processing. So the recovery and reuse of water can reduces the environmental pollution and quantity of effluents and rises supplement water resources (Saxena et al. 2017).

4.10 Electrochemical Dyeing

The vat and Sulphur dyeing processes are having both reducing and oxidized step. Dyes used in the process are water insoluble, which made into water-soluble form by the addition reducing agents and alkali. These reducing agents are producing the non-regenerable oxidized byproduct. These conventional dyeing processes are non-ecofriendly in nature because of the discharge of waste effluent from dyeing. Electrochemical method is an alternative technique to overcome those problems from the conventional dyeing process but still in the laboratory stage (Sarita 2016).

4.11 Foam Finishing

In foam finishing process, air is widely used instead of water in the form of dispersion foam, which leads to the energy conservation. The mechanical air blowing process using excess stirring and the mixture of chemical or dyes and foaming agents' forms foam mechanically. Foam is coated on the surface of the fabric and then this foam-coated fabric is placed between the squeeze rolls to collapse the foam and uniformly distributed the chemical or dye on the fabrics. Blow ratio chosen the relative proportion of air and liquid phase. The significant parameters for the foaming are density, diameter and foam stability need to remain constant. In general, the foam density for foam finishing operation is in the range of 0.14 g/cct–0.07 g/cc and 0.33–0.20 g/cc for foam printing. The foam density is used in process is mainly based on the fabric weight which increased by increasing the fabric weight. Bubble diameter used in the foam processing is in the range of 0.0001–2.0 mm (Sarita 2016). When comparing the large size bubbles, smaller size bubbles are more stable. This technique is applied for fabric preparation, printing, dyeing, softening water, oil repellent finish, DP finish, soil-release finish, and mercerization, etc.

5 Detoxifying Supply Chain

In the 2011, Greenpeace (non-governmental environmental organization) reported that the textile industries in china releasing hazardous chemicals into rivers. The organization wrote the open letter to the multinational fashion brands to "detox" their manufacturing process. The main of the goal of the organization is to remove 11 most toxic chemicals from the manufacturing process by 2020 and communicate clearly about the outsourced production suppliers of the companies (Ethical fashion form 2013). In 2011, the detox campaign was launched for presenting the dirty laundry report about the usage of toxic chemicals in textile industries like China and Mexico. Based on the report, Greenpeace conduct test for toxic chemicals across 20 global brands. The toxic chemicals from the industry like nonylphenolethoxylates (NPEs) in textile that is very toxic and durable chemical (Ethical fashion form 2013). They have ability to contaminate the water bodies and food chain. The needs of the Detox campaign are

- Major fashion brands to accept the zero chemical discharge
- Brands and supplier transparency across their manufacturing process
- Consumer's conversation with the brands and government to restrict the import and sales of the toxic chemical containing goods.

Stakeholders from chemical and fashion industries have to be dedicated to following international regulations for the production the toxic free cloth. The non-government organization states the whole supply chain try to meet the environmental demand. For the sustainable production and configuration with Greenpeace DETOX commitment, comprehensive services, training and workshop must to conduct. Apparel brands and supplier have joined to remove all hormone-disrupting, toxic and persistent chemical from the products and processes. In the Greenpeace campaign, the top brands and retailer are working for achieving a "Zero discharge of hazardous chemicals" (ZDHC) in textile supply chain by the year 2020. The motive of this ZDHC is to create a Joint Roadmap for eliminating the 11 toxic chemicals and helping the member companies to replace the greener chemicals in manufacturing process. The current consumers are expecting safety standards and liable production processes to protect the environment and product from harmful chemicals.

A chemical management system is implemented to verify and monitor process that is interconnected to the manufacture, distribution and sale of product. The process is economical owing to the lesser consumption of water and energy for the treatment of waste.

Experts conduct full inspection of the supply chain and then give the clear solution to the textile and footwear industries. The main aim of this service is to diminish the negative environment impact compliance and increase compliance that is associated with the industrial operation and practices. The services are as followed:

- Provide proper flexible module-based training to hazardous substance, chemical and environment management, action plan workshop.
- Monitoring of wastewater for hazardous substance based on the guideline of ZDHC's wastewater and DETOX commitment.
- Give Manufacturing restricted substance list (MRSL) compliance to the input chemical inventory.
- Supplier audits according to Zero discharge of hazardous chemicals chemical management system (ZDHC CMS).
- Product testing based on the standards, regulations and DETOX commitment.
- Water efficiency program, risk assessment for Continuous Improvement through.
- A customized chemical management audits for evaluating the procurement and the raw materials and chemicals storage.
- Verification audits for the control and disposal of pollutants and validate corrective action (Sharma 2016).

These are the services helpful to improve the performance of the supply chains in textile industries.

5.1 Supply Chain Management

A supply chain is a network between a company and their supplier to produce an end product and deliver to customer. System of organization, people, activities, information, and resources are involved in the supply chain. The supply chain management improves the supply chain. Supply chain management (SCM) is management of raw materials, information and finance as they move in the way from supplier to manufacturer to wholesaler to retailer to consumer. It is important to the company success as well as the customer satisfaction. The goal of the supply chain management (SCM) is waste reduction, time compression, flexible response and unit cost reduction.

The flow of the supply chain management is classified into three types and is product flow, information flow and finances flow. The product flow comprises a smooth flow of textiles from suppliers to customer besides service needs. The quicker material flow is essential for the enterprise simultaneously it minimize the cash cycle. Information flow includes the request for quotation, transmitting order and updating of delivery status updates. The financial flow is also called as money flow, which consists of the payment schedule, credit terms, and consignment arrangements. The management of all three flows is important to attain an efficient and effective supply chain (https://en.wikipedia.org/wiki/Supply_chain_management).

Advantages of supply chain management are explained as follows:

- Improves better customer services and relationship.
- Reduces warehouse and transportation costs.
- Reduces direct and indirect costs.

- Makes well delivery mechanism for products and services in demand with minimum delay.
- Develops the productivity and business functions.
- Helps to achieve shipping of the right products to the right place at the right time.
- Improves inventory management and support the successful execution of just-in-time stock models.
- Supports companies to reduce waste, driving cost and attaining efficiency in the supply chain process.
- Assists companies in adjusting to the challenges globalization, expanding consumer expectations and economic upheavals.

5.2   Supply Chain Decision

The decisions for supply chain management can be generally divided into three levels such as strategic, tactical, and operational.

5.2.1   Strategic Decision

Company management creates high-level strategic supply chain decisions, which is related to entire organizations. The strategic decisions are closely connected to the corporative strategy and guide supply chain policies from a design perspective. These comprise product development, customers, suppliers, and logistics. In the company, senior management has to define a strategic method for the manufacturing and selling of product to their consumers. When product cycle mature or product sales decline, management needs to take strategic decision to develop and introduce new forms of existing products into the market. A company has to recognize consumers for their products and services at the strategic level. They have to identify the key customer segments where company marketing and advertising will be targeted, when management takes strategic decisions on the products to manufacture. Manufacturing decision explain requirement of manufacturing infrastructure and technology. The company management needs to create strategic decision on in what way product to be manufactured. The manufacturing decision requires new manufacturing facilities to be built or to improve production. With respect to suppliers, the company needs to decide on the strategic supply chain policies. Minimize the purchasing spend for a company that can be related to increase profit and the number of decisions can be made to get that result. The logistics function is important to the success of the supply chain. Order fulfillment is a significant part of the supply chain and company management has to make the strategic decisions on the logistics network. The performance of the supply chain is depends on the design and operation of the network (Murray 2016).

5.2.2 Operational Decision

Operational decisions are made at the business locations that affect in what way product are developed, sold, moved and manufactured. They give awareness to the strategic and tactical decisions that have been approved in a company. A framework inside the company's supply chain operations is created by these higher-level decisions. These decisions make sure that the product efficiently moves along the supply chain and achieving the maximum cost benefit. The local plant management has to define an operational decision for keeping certain products in stock to confirm that manufacturing is to be continued. The local management is important to define the operational decision to negotiate with supplier for them to make a product with better to confirm the quality of the finished product. Even though strategic and tactical decisions are made to get the highest efficiency with low cost, the day-to-day operations of supply chain need that local management makes hundreds of operational decisions. These decisions are made in the framework that created by the strategic and tactical processes however that not made in isolation (Murray 2016).

5.2.3 Tactical Decision

Tactical decision made to focus on adopting measures, which will produce cost benefit for a company. These decisions enclose the extent of the supply chain for a whole company and creating real benefits for the company. Company management may be made strategic decisions on the number and location of manufacturing places to be operated. However, tactical decisions are made on how to produce product within the budget and made on adopting of manufacturing methodologies like just-in-time. These decisions made to reduce material wastage in a company still cannot distribute to other manufacturing plants. Tactical decision may be needed to use a third party logistics company in country where transportation costs are high. At a tactical level, management needs to work in strategic guidelines to detect and negotiate terms, which will understand the maximum cost benefits across a company. A tactical decision has to make for developing the particular product (Murray 2016).

5.3 Challenges and Solutions in Sustainable Wet Processing

During the wet processing in the textile industries, several problems have been occurred in each stage of wet processing such as preparation, dyeing and finishing that affect the quality of fabric material and effluents discharge. The water and chemicals are very important for the textile processing and are a significant part of production costs. Water is flowing through different supply sources that have impurities in some amount. The impurity level is based on the supply source and the contaminants comprises of heavy metals, calcium, magnesium, aluminum, chlorine, dissolved salt, oil and grease etc., which influence the poor wet processing operations that also damage the machineries used. The problem occurred by these impurities in the water in the wet processing operation such as

- The poor removal starch sizes during desizing.
- Irregular absorbency after scouring and tendency of fabric material to attract soil.
- Reduction of surfactants solubility and rate of dissolution.
- Fiber degradation, decrease of whiteness, loss of fiber strength occurred during the catalytic decomposition of hydrogen peroxide.
- In dyeing, inconsistent shade and spots because of uneven washing off.
- Reduction in the wet fastness and foaming.
- Corrosion of machinery.

Therefore, the proper water treatment is necessary and the addition of suitable sequestrants during wet processing operation also to overcome these problems.

For each stages of wet processing, more amounts of chemicals are spent to improve the quality of textile material and also improve the processing operations. The production cost increased by increasing the chemical consumption. For saving the production cost, recycle and reuse of chemicals are essential. The usage the chemicals in the textile processing can affect the effluents that released from each stage of operations that increases risks the agricultural pollution, disrupt the aquatic ecosystem and hazardous to the humans. Thus, the lesser dangerous chemicals or bio-based dyes/chemicals can be used instead of the potentially hazardous and carcinogenic chemicals/dye to improve sustainability of the wet processing operation. The certain of chemical substitutions have been explained as follows:

- The biological oxygen demand in the effluent can be reduced by the substitution of formic acid for acetic acid.
- Chlorine can be used instead peroxide compounds in bleaching to remove the significant quantity of activated carbon absorbable organohalogen and trichloromethane compounds.
- The sodium hydroxide or sodium carbonate buffer can be substituted for trisodium phosphate to eliminate the phosphate pollution. This trisodium is mainly used in the fixation of high temperature reactive dyes. The effluent discharged after dyeing may have lesser salt content so the pollution of the rivers can be controlled (Schlaeppi 1998).
- Synthetics dyes can be replaced by bio based dyes and natural extracts to reduce the toxicity of the effluents.

6 Conclusion

In this chapter, the sustainable wet processing operation has been discussed with the recent innovations and techniques such as plasma technology, supercritical fluid technology, ultrasonic waves, and digital ink-jet printing and so on. But, the new techniques are still facing challenges even in the laboratory level which requires further development and improvement to make it as an effective substitute to the conventional processing. The production of natural and bio-based dyes and chemicals are essential for the sustainable wet processing to control the environmental problems. The developing countries are in need to concentrate on the installation of the proper waste treatment units, and disposal. Thus, the governments need to impose the regulations for the textile industries to produce sustainable textile materials. The main aim of the textile industries is to achieve the zero discharge of hazardous chemicals that prevent the humans and aquatic life of environment. Thus, the manufacturer, retailer and consumer are essential to be conscious about environmental problems and need to produce the toxic free products.

References

Anastas PT, Warner JC (1998) Green chemistry: theory and practice. Oxford University Press, Oxford (England), New York. ISBN 9780198502340

Anon (2015) 30 shocking figures and facts about textile and apparelindustry. www.business2 community.com/fashion-beauty/30-shocking-figures-facts-global-textileapparel-industry-0122 2057#ADLVmKPEeCsdUlzQ.99. Accessed 8 Mar 2016

Arputharaj A, Raja ASM, Sexena S (2015) Developments in sustainable chemical processing of textiles. Green fashion, Singapore, pp 217–252

Beckmann W, Pflug J (1983) Reuse of weakly loaded liquiors from textile processing operations. Textil-Praxis Int 38:II–VI

Cegrra J, Puente P, Valldeperas J (1992) The dyeing of textile materials, the scientific bases and techniques of application. Translated from Spanish by instituto per la textilia. Nuova oflito, Torino

Correia VM, Stephenson T, Simon JJ (1994) Characterisation of textile waste water—a review. Environ Technol 15:916–929

Eccles S (2016) What you should know about digital textile inks. https://www.fespa.com/item/5467-what-you-should-know-about-digital-textile-inks.html. Accessed 23 May 2016

Economic Times (2014) India world's second largest textiles exporter: UN Comtrade (June 2)

Ethical Fashion Form (2013) Fashion detox: reducing harmful chemicals in supply chains. http://source.ethicalfashionforum.com/digital/the-bigfashion-detox-reducing-harmful-chemicals-in-fashion-supply-chains. Accessed 10 Apr 2013

Ibrahim DF (2012) Clean trends in textile wet processing. J Textile Sci Eng 2:e106. doi:10.4172/2165-8064.1000e106

Jones HR (1973) Pollution control in the textile industry: Noyes Data Corporation. Park Ridge, New Jersey, p 323

Murray M (2016) Logistics and supply chain management. https://www.thebalance.com/operational-supply-chain-management-2221189. Accessed 18 Nov 2016

O Ecotextiles (2009) What is the energy profile of the textile industry? http://oecotextiles.wordpress.com/2009/06/16/what-is-the-energy-profile-of-the-textile-industry/. Accessed 30 Apr 2010

Osman AT (2007) Analysis of industrial wastewater, case study: textile industry (Msc), Sudan academy of science

Parisi ML, Fatarella Spinelli D et al (2015) Environmental impact assessment of an eco-efficient production for coloured textiles. J Clean Prod 108:514–524

Ramesh Babu B, Parande AK et al (2007) Cotton textile processing: waste generation and effluent treatment. J Cotton Sci 11:141–153

Rott U, Minke R (1999) Overview of wastewater treatment and recycling in the textile processing industry. Water Sci Technol 40:37–144

Sarita S (2016) Recent developments in textile wet processing. http://www.indiantextilejournal.com/News.aspx?nId=0IQyeFv+d//Gbr/dwPp82g==. Accessed Feb 2016

Saxena S, Raja ASM, Arputharaj A (2017) Challenges in sustainable wet processing of textiles. In: Muthu SS (ed) Textiles and clothing sustainability, textile science and clothing technology. doi:10.1007/978-981-10-2185_2

Schlaeppi F (1998) Optimizing textile wet processes to reduce environmental impact. Text Chem Colorist 30:19–26

Sharma G (2016) Fashion from alliance to detox manufacturing process. Network for business sustainability. Accessed 25 Jan 2016

Shukla SR (2007) Pollution abatement and waste minimisation in textile dyeing. In: Christie RM (ed) Environmental aspects of textile dyeing. Woodhead Publishing Limited, Cambride

Smith B (1986) Identification and reduction of pollution sources in textile wet processing. Pollution prevention program, North Carolina Department of Environment, Health and Natural Resources, Raleigh, North Carolina, p 129

Green Dyeing of Cotton- New Strategies to Replace Toxic Metallic Salts

Shahid-ul-Islam and B.S. Butola

Abstract Plant extracts have acquired tremendous commercial potential for their use in textile dyeing and finishing applications instead of toxic synthetic dyes which produce hazardous chemicals. Synthetic dyes particularly azo dyes have been banned in many countries. On the other hand, polymer and textile scientists have focussed their momentum to investigate dyeing chemicals and auxiliaries from plant materials. Phytochemicals including colorants from plant extracts exhibit many advantageous functions such as better biodegradability, higher compatibility with the environment, lower toxicity and also produce soft and sober shades. Among different fibres cotton has been extensively dyed with colorants derived from plant sources such as Weld (*Reseda luteola*), Annato (*Bixa orellana*), Sage (*Salvia officinalis*), Henna (*Lawsonia inermis*), Saffron (*Crocus sativus*), Madder (*Rubia tinctorum, R. cordifolia*), and Sappanwood (*Caesalpinia sappan*). In this chapter we review some of the commonly used plant extracts in cotton dyeing. This chapter also describes latest research that is underway in this area to replace metal salt mordants which create pollution and human health problems.

Keywords Metal salts · Dyes · Madder · Biomordants · Enzymes

1 Introduction

Color is one of the most important attributes of clothing, being considered as a quality indicator and determining frequently their acceptance in today's demanding consumer market place (Yusuf et al. 2015; Zhang et al. 2014a; Zhao et al. 2014). Owing to their many advantages, natural dyes and pigments from plants, animals and minerals have found various applications in our day-to-day life including food (Sivakumar et al. 2009) and pharmaceutical preparation, cosmetic products, textile coloration (Shahid et al. 2013; Yusuf et al. 2016a, 2013), energy harvesting

Shahid-ul-Islam (✉) · B.S. Butola
Department of Textile Technology, Indian Institute of Technology, New Delhi 110016, India
e-mail: shads.jmi@gmail.com

© Springer Nature Singapore Pte Ltd. 2018
S.S. Muthu (ed.), *Detox Fashion*, Textile Science and Clothing Technology,
DOI 10.1007/978-981-10-4876-0_3

systems, solar protectors, UV and thermo-sensors. Many synthetic dyes on the other hand have been banned by many countries because of the possible toxic and harmful effects associated with their use (Rather et al. 2015, 2016a; Vankar et al. 2006). The demand for natural dyes, such as carotenoids, curcuminoids, anthraquinones, anthocyanins and betalains has increased recently due to the their environmental friendly nature coupled with global trend of maintaining good health and reducing the risk of disease (Islam et al. 2014a; Vankar et al. 2007; Vankar and Shukla 2012). Over the past few decades, different natural fabrics have been colored with plants extracts. Among them the use of cotton predates recorded history (Zhou and Kan 2014). Research by archaeologists indicates its use in clothing in 3000 BC. Cotton is a warm weather shrub cultivated in America, India, China, and Egypt Africa. It grows in a ball, or protective capsule, around the seeds of plant species belonging to the genus *Gossypium* (Islam and Mohammad 2016b). The different species of this genus include *Gossypium hirsute, Gossypium barbadense, Gossypium arboretum,* and *Gossypium arboretum.* Cotton is mainly composed of cellulose, along with other minor compounds such as waxes, monomeric and polymeric sugars, residual protoplasm and minerals. Roadmap of cotton dyeing is summarized in Fig. 1. The main plants along with their dyeing principles which have been used to dye cotton are described below.

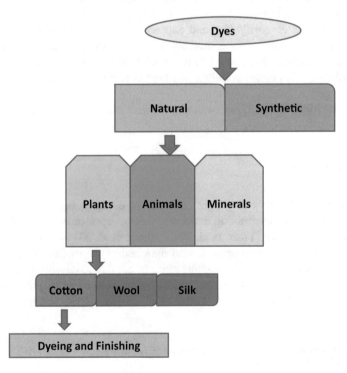

Fig. 1 Roadmap of cotton dyeing with plant extracts

2 Plant Extracts

2.1 *Rubia Tinctorum*

Madder is one of the most important dye plant native to Western and Central Asia and Mediterranean region (Mayer and Cook 1943; Perkin and Everest 1918). Over the past few decades, the health benefits of *Rubia tinctorum* have received much interest (Mikropoulou et al. 2009; Shahmoradi Ghaheh et al. 2014). It is well documented that all parts of *Rubia tinctorum* have found use in treatment of various human diseases. Madder dyes are well known ancient red dyes extracted from the grounded roots of *Rubia tinctorum* (Gupta et al. 2001a; Van Stralen 1993; Yusuf et al. 2013). The main dyeing principles present in the madder are hydroxyl anthraquinones including alizarin, purpurin, pseudopurpurin, rubiadin, munjistin, xanthopurpurin, lucidin and ruberythric acid (Gupta et al. 2001b; Vera de Rosso and Mercadante 2009). Alizarin is the major dye compound present in *Rubua tinctorum*. The chemical structures of dyeing compounds are shown in Fig. 2. Madder contain carbonyl and hydroxyl functional groups and are classified as mordant dyes. They have extensively been used to produce a variety of hues with good fatness properties and other functional properties such as antimicrobial and UV protection on cotton, silk, wool and synthetic fibres in conjunction with number of different mordants (El-Shishtawy et al. 2009; Farizadeh et al. 2009; Parvinzadeh 2007).

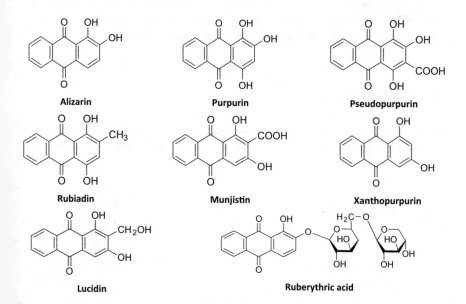

Fig. 2 Chemical compounds present in *Rubia tinctorum* root extracts

2.2   *Rubia cordifolia*

Rubia cordifolia commonly known as Indian madder belongs to the family *Rubiaceae*. It is found in South East Asian countries India, Srilanka, Malaysia, Japan and also in Tropical Africa (Perkin and Everest 1918; Yusuf et al. 2015). It is widely grown throughout Tropical Asia for its excellent red dye. The chief coloring component present in the roots is a mixture of purpurin (C.I. 75410), pseudopurpurin (C.I. 75420), munjistin (C.I. 75370), xanthopurpurin (C.I. 75340) and nordamn-canthal (Fig. 3) (Bhuyan and Saikia 2005; Yusuf et al. 2017). This dye is used in the coloring of different textile substrates including cotton (Bechtold and Mussak 2009). The alizarin dyeing compound is missing in *Rubia cordifolia*. Pharmacological research has also shown that *Rubia cordifolia* has been used in traditional medicine and shows wide range of biological properties (Boldizsár et al. 2006; Islam and Mohammad 2016a).

2.3   *Rubia Sikkimensis*

Rubia sikkimensis is a perennial branched climber belonging to the family Rubiaceae is well recognised medicinal plant (Anonymous 1972; Islam et al. 2016). It is distributed from Nepal eastwards to Assam, Nagaland and Manipur. It is closely related to *Rubia cardofolia* and is the main dye-yielding plant in North Eastern India (Gulrajani and Gupta 1992; Islam and Mohammad 2015a, b). The red dye present in its roots has been used by hill tribes for dyeing clothes, human hair, decorations for spear and ornaments, cane and bamboo articles. The main dyeing component is purpurin along with munjistin, xanthopurprin as minor compounds (Perkin and Everest 1918).

2.4   *Rheum emodi*

Rheum emodi also known as Indian Rhubarb belongs to family *Polygonaceae* has been used in traditional medicine since ancient times. *Rheum emodi* is distributed in

Fig. 3 Chemical compound present in *Rheum cardifolia* extracts

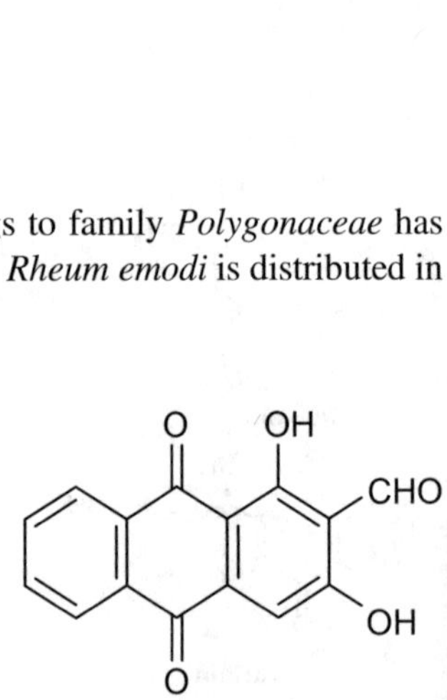

Nordamncanthal

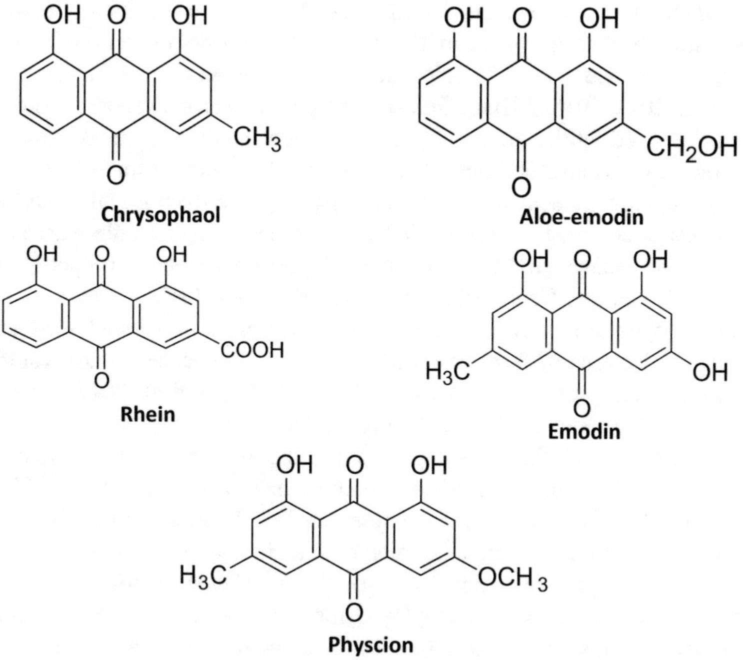

Fig. 4 Chemical compounds present in *Rheum emodi* root extract

temperate and sub tropical regions found in Himalayas from Kashmir to Sikkim and also in Assam (Perkin and Everest 1918). Indian rhubarb roots contain a large number of phytochemicals mainly anthraquinones and their derivatives (free as well as glycosides) (Gupta and Gulrajani 1996). The chemical isolated from Indian rhubarb roots include chrysophaol, aloe-emodin, rhein, emodin and physcion (Fig. 4). These anthraquinones have shown a wide range of pharmacological activities including cathartic, antibacterial, diuretic, anticancer, hepatoprotective, anti-inflammatory, and analgesic effects (Agarwal et al. 2000; Nayak 2014). Indian Rhubarb colorants are widely used in dyeing and functional finishing of textiles (Das et al. 2008; Vankar et al. 2007; Zhou et al. 2014).

3 New Approaches to Replace Toxic Mordants

Cotton has less affinity with phtochemicals present in plant extracts and therefore results in narrow range of shades, and low colour fastness results (Rather et al. 2016b; Vankar et al. 2007; Yaman et al. 2009). To overcome this problem, several metal salts as mordants have been investigated by several researchers.

Various metallic salts of Fe, Cu, Sn, Mg, and Pb have been frequently used and it has been noticed that the use of mordants not only enhances dye fibre interactions but may also induce change in colorimetric properties in terms of CIELab values (Yusuf et al. 2012, 2015, 2016c). Colorimetric parameters and fastness properties of the naturally dyed fabrics are greatly dependent on the type of metal salt and their metal complex forming abilities (Tsatsaroni et al. 1998b). Mordanting methods including pre-simultaneous and post generally improve dyeing ability and enable the chemists or colourist to produce hues on a wide range of textile materials with expanded shade ranges and better color wash and light fastness properties (Khan et al. 2012; Tutak and Benli 2011; Tutak and Korkmaz 2012).

Mordant dyes are known to form a complex in which metal ions can act acceptors of electrons to form co-ordinate bonds with dye molecules resulting in improved fastness properties (Zhang et al. 2014b; Zhao et al. 2014; Zhou et al. 2014). It is well established fact that mordant dye complex may be formed by first pre-mordanting, by simultaneous application of the metal salt and the dye called as simultaneous mordanting or by post-mordanting method (Shahid et al. 2013; Islam and Mohammad 2015). A number of research investigations are available in literature on different types of metallic salts, mordanting methods and their role in the development of elegant hues and increased dye uptakes on wool, cotton and silk fabrics. Some of the commonly employed mordants in natural dyeing technology include metallic salts such as aluminium potassium sulfate, potassium dichromate, stannous chloride (falling under brightening mordant category), ferrous sulphate and copper sulfate (falling under dulling mordant category), tannic acid and oil mordants (Rather et al. 2015, 2016b). Considerable research work has been undertaken around the globe on the use of metallic salt mordants and satisfactory results have been obtained with cotton fabrics on lab scale (Islam et al. 2014b; Rather et al. 2016c). Most of the mordants are mainly from p and d-block of periodic table however rare earth chlorides from s-block have also displayed good results on ramie fabrics (Zheng et al. 2011) (Table 1). Adeel et al. (2009) in a research experiment studied the use of Cu (copper sulphate), Al (aluminium sulphate), Fe (iron sulphate) and tannic acid as mordants in the dyeing of cotton previously irradiated using different dosses of gamma radiations with pomegranate dye. Likewise, Batool et al. (2013) employed 4% ferrous sulfate as pre-mordant and tannic acid as post-mordant and investigated their role in the development of colorful cotton with chicken gizzard leaves powder. Rehman et al. (2012) studied the application of lawsone dye extracted from henna leaves on cotton previously irradiated with different doses of gamma radiations using copper and iron as mordants. All these experiments have shown that selected metal mordants form stable complexes with dye molecules and results in enhanced functional properties.

On one hand metallic salts are known as wonder agents however on the other side are known to cause serious human health and environmental challenges. To mitigate this pollution in cotton industry, the use of biomordants, enzymes, and biopolymers in place of toxic metals allows sustainable cotton dyeing and finishing. The schematic view of different agents used in coloration of cotton is shown in Fig. 5. Natural mordanting involves application of tannins which are classified as water soluble

Table 1 Commonly employed metal salts in cotton coloring (modified from Shahid et al. 2013)

Mordants	Group
• Alum	s-block
• Ferrous sulfate	d-block
• Stannous chloride	p-block
• Copper sulfate	d-block
• Potassium dichromate	s-block
• Stannous sulphate	p-block
• Cobalt sulphate	d-block
• Aluminum sulphate	p-block
• Magnesium sulphate	s-block
• Zinc sulphate	d-block
• Manganese sulphate	d-block
• Nickel sulphate	d-block
• Stannic chloride	p-block
• Ferric chloride	d-block
• Aluminum chloride	p-block
• Copper chloride	d-block
• Zinc Chloride	d-block
• Calcium hydroxide	d-block
• Zinc tetrafluoroborate	d-block

phytochemicals found in wide variety of plant species. Tannin mordants have been derived from a number of plant species including *Quercus ithaburensis* ssp. *macrolepis, Punica granatum* L., *Rosmarinus officinalis, Thuja orientalis* (Islam et al. 2014a, b), *Quercus infectoria* (Yusuf et al. 2016b) *Eurya acuminate* DC var *euprista* Karth (Vankar et al. 2008), *Acacia catechu* (Mansour et al. 2016; Yusuf et al. 2017), *Tamarindus indica* L. (Prabhu and Teli 2014), *Emblica officinalis* (Prabhu et al. 2011), *Terminalia chebla* (Samantaa and Agarwal 2009), *Rumex hymenosepolus* (Haji 2010), chlorophyll (Guesmi et al. 2013), and have been effective in crosslinking cotton and other fabrics with natural dyes.

Several researches have been conducted on dyeing and functional finishing of cotton with natural dyes using softer enzymes such as alpha-amylase, amyloglycosidase and trypsin. These enzymes have the ability to increase dye uptake and other functional properties. Tsatsaroni et al. (1998a) assessed the color and fastness of cotton and wool fabrics dyed with chlorophyll and carmine pigments previously treated with cellulase, alpha-amylase and trypsin enzymes and noticed that the use of enzymes improved wash and light fastness properties. Liakopoulou-Kyriakides et al. (1998), likewise obtained good dyeing results with the application of alpha-amylase and trypsin enzymes on cotton and wool dyed with *Crocus sativus* dye. Based on a research work administered by Vankar and Shankar (2008), the sustainable ultrasonic coloration of cotton with *Acacia catechu* and *Tectona grandis* natural dyes was studied in the presence and absence of protease; α-amylase; lipase and diasterase enzymes. They reported that the pretreatment of cotton with these enzymes

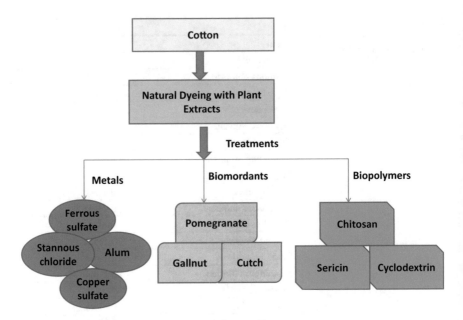

Fig. 5 Various agents employed in cotton dyeing and finishing

enhances colorimetric values besides improving wash and light fastness properties. Vankar and co workers also introduced enzyme pretreatment of protease-amylase, diasterase and lipase in order to increase dyeing potential of three natural dyes *Terminalia arjuna*, *Punica granatum*, and *Rheum emodi* onto cotton and silk. It was observed that the tannic acid-enzyme-dye combination showed much better colorimetric and fastness properties and offer softer chemistry option to metal mordants (Vankar et al. 2007). A more detailed discussion about the use of enzymes in cotton fabrics is discussed in a recent review published by Shahid et al. (2016).

Apart from enzymes biopolymers such as chitosan, sericin, and cyclodextrin have been applied as pre-treatment agents in most current research studies. Kim (2006) in a research work observed that chitosan when used as mordant on cotton dyed with green tea extract was found to improve dyeing characteristics and the UV protection property. In another study adsorption and thermodynamic studies of lac dyeing on cotton pretreated with chitosan was investigated by Rattanaphani et al. (2007). The dyeing of cotton was performed under optimum natural dyeing conditions of pH 3, a material to liquor ratio of 1:100 and a contact time of 3 h, and it was observed that chitosan treatment significantly enhances dye adsorption onto cotton. Chairat et al. (2008) also described kinetic parameters of lac dyeing on cotton pretreated with chitosan. They used pseudo first- and second-order kinetic models to examine the fitting of experimental data and found that pseudo second-order kinetic model with an activation energy of 42.4 kJ/mol gave the best fitting. Other polymers are deeply discussed in one of the recent review paper published by Islam et al. (2013).

4 Conclusion

Over the past few decades much research work has been done using renewable resources as source of dyes for functional finishing of cotton and other textile materials. This is due to the fact that natural colorants cause minimum pollution and having less risk associated with their use. Extensive research is available on dyeing of cotton with plant extracts using different metallic salts as mordants. Although mordants are well known to produce wide range of shades but unfortunately are associated with many ricks including environmental and human health issues. Keeping this in view, the use of low-environmental impact technologies involving use of polyphenolic biomordants, enzymes and sustainable biopolymers offers full potential to clean up cotton dyeing industry in the near future.

Acknowledgment Dr Shahid-ul-Islam gratefully acknowledge financial support from the DST-SERB through National Postdoc Fellowship (Grant no. PDF/2016/003859).

References

Adeel S, Ali S, Bhatti IA, Zsila F (2009) Dyeing of cotton fabric using pomegranate (Punica granatum) aqueous extract. Asian J Chem 21:3493–3499

Agarwal SK, Singh SS, Verma S, Kumar S (2000) Antifungal activity of anthraquinone derivatives from Rheum emodi. J Ethnopharmacol 72:43–46

Anonymous (1972) Wealth of India: raw materials vol IX. CSIR, New Delhi

Batool F, Adeel S, Azeem M, Ahmad Khan A, Ahmad Bhatti I, Ghaffar A, Iqbal N (2013) Gamma radiations induced improvement in dyeing properties and colorfastness of cotton fabrics dyed with chicken gizzard leaves extracts. Radiat Phys Chem 89:33–37

Bechtold T, Mussak R (2009) Handbook of natural colorants. Wiley, Chichester, UK

Bhuyan R, Saikia CN (2005) Isolation of colour components from native dye-bearing plants in northeastern India. Bioresour Technol 96:363–372

Boldizsár I, Szűcs Z, Füzfai Z, Molnár-Perl I (2006) Identification and quantification of the constituents of madder root by gas chromatography and high-performance liquid chromatography. J Chromatogr A 1133:259–274

Chairat M, Rattanaphani S, Bremner JB, Rattanaphani V (2008) Adsorption kinetic study of lac dyeing on cotton. Dyes Pigm 76:435–439

Das D, Maulik SR, Bhattacharya SC (2008) Colouration of wool and silk with Rheum emodi. Indian J Fibre Text Res 33:163–170

El-Shishtawy RM, Shokry GM, Ahmed NSE, Kamel MM (2009) Dyeing of modified acrylic fibers with curcumin and madder natural dyes. Fibers Polym 10:617–624

Erdem İşmal Ö, Yıldırım L, Özdoğan E (2014a) Use of almond shell extracts plus biomordants as effective textile dye. J Clean Prod 70:61–67

Erdem İşmal Ö, Yıldırım L, Özdoğan E (2014b) Valorisation of almond shell waste in ultrasonic biomordanted dyeing: alternatives to metallic mordants. J Text Institute 1–11

Farizadeh K, Montazer M, Yazdanshenas ME, Rashidi A, Malek RMA (2009) Extraction, identification and sorption studies of dyes from madder on wool. J Appl Polym Sci 113:3799–3808

Guesmi A, Ladhari N, Hamadi NB, Msaddek M, Sakli F (2013) First application of chlorophyll-a as biomordant: sonicator dyeing of wool with betanin dye. J Clean Prod 39:97–104

Gulrajani ML, Gupta D (1992) Natural dyes and their application to textiles. Department of Textile Technology, IIT Delhi, New Delhi, India

Gupta D, Kumari S, Gulrajani M (2001a) Dyeing studies with hydroxyanthraquinones extracted from Indian madder. Part 1: dyeing of nylon with purpurin. Color Technol 117:328–332

Gupta D, Kumari S, Gulrajani M (2001b) Dyeing studies with hydroxyanthraquinones extracted from Indian madder. Part 2: dyeing of nylon and polyester with nordamncanthal. Color Technol 117:333–336

Gupta DB, Gulrajani M (1996) The light fading mechanism of dyes derived from rhubarb extract. J Soc Dyers Colour 112:269–272

Haji A (2010) Functional dyeing of wool with natural dye extracted from Berberis vulgaris wood and Rumex hymenosepolus root as biomordant. Iran J Chem Chem Eng 29

Islam S, Mohammad F (2015a) High-energy radiation induced sustainable coloration and functional finishing of textile materials. Ind Eng Chem Res 54:3727–3745

Islam S, Mohammad F (2015b) Natural colorants in the presence of anchors so-called mordants as promising coloring and antimicrobial agents for textile materials. ACS Sustain Chem Eng 3:2361–2375

Islam S, Mohammad F (2016a) Potent polyphenolic natural colorants derived from plants as eco-friendly raw materials for the dyeing industry. In: Muthu SS, Gardetti MA (eds) Green fashion, vol 2. Springer Singapore, Singapore, pp 229–247

Islam S, Mohammad F (2016b) Sustainable natural fibres from animals, plants and agroindustrial wastes—an overview. In: Muthu SS, Gardetti M (eds) Sustainable fibres for fashion industry, vol 2. Springer Singapore, Singapore, pp 31–44

Islam S, Shahid M, Mohammad F (2014a) Future prospects of phytosynthesized transition metal nanoparticles as novel functional agents for textiles. In: Tiwari A, Syvajarvi M (eds) Advanced materials for agriculture, food, and environmental safety. John Wiley & Sons, Inc., pp 265–290

Islam S, Rather LJ, Shahid M, Khan MA, Mohammad F (2014b) Study the effect of ammonia post-treatment on color characteristics of annatto-dyed textile substrate using reflectance spectrophotometery. Ind Crop Product 59:337–342

Islam S, Shahid M, Mohammad F (2013) Green chemistry approaches to develop antimicrobial textiles based on sustainable biopolymers-a review. Ind Eng Chem Res 57

Islam S, Rather LJ, Shabbir M, Bukhari MN, Shahid M, Ali Khan M, Mohammad F (2016) Bi and tri metal salt combinations plus colorants extracted from timber industry waste as effective dyeing materials to produce novel shades on wool. J Nat Fibers 1–11

Khan SA, Ahmad A, Khan MI, Yusuf M, Shahid M, Manzoor N, Mohammad F (2012) Antimicrobial activity of wool yarn dyed with Rheum emodi L. (Indian Rhubarb). Dyes Pigm 95:206–214

Kim S-H (2006) Dyeing characteristics and UV protection property of green tea dyed cotton fabrics. Fibers Polym 7:255–261

Liakopoulou-Kyriakides M, Tsatsaroni E, Laderos P, Georgiadou K (1998) Dyeing of cotton and wool fibres with pigments from Crocus sativus—effect of enzymatic treatment. Dyes Pigm 36:215–221

Mansour R, Mighri Z, Mhenni F (2016) Exploring the potential uses of Vitis vinifera L. leaves as raw material for textile dyeing without metal mordants. Fibers Polym 17:1621–1626

Mayer F, Cook AH (1943) The chemistry of natural coloring matters: the constitutions, properties, and biological relations of the important natural pigments. Reinhold publishing corporation, New York

Mikropoulou E, Tsatsaroni E, Varella EA (2009) Revival of traditional European dyeing techniques yellow and red colorants. J Cultural Heritage 10:447–457

Nayak L (2014) A study on coloring properties of Rheum emodi on jute union fabrics. J Text 2014:1

Parvinzadeh M (2007) Effect of proteolytic enzyme on dyeing of wool with madder. Enzym Microb Technol 40:1719–1722

Perkin AG, Everest AE (1918) The natural organic colouring matters. Longmans, Green and Co., London

Prabhu KH, Teli MD (2014) Eco-dyeing using *Tamarindus indica* L. seed coat tannin as a natural mordant for textiles with antibacterial activity. J. Saudi Chem. Soc. 18:864–872

Prabhu KH, Teli MD, Waghmare NG (2011) Eco-friendly dyeing using natural mordant extracted from *Emblica officinalis* G. Fruit on cotton and silk fabrics with antibacterial activity. Fibers and Polym 12:753

Rather LJ, Islam S, Azam M, Shabbir M, Bukhari MN, Shahid M, Khan MA, Rizwanul Haque QM, Mohammad F (2016a) Antimicrobial and fluorescence finishing of woolen yarn with *Terminalia arjuna* natural dye as an ecofriendly substitute to synthetic antibacterial agents. RSC Advan 6:39080–39094

Rather LJ, Islam S, Khan MA, Mohammad F (2016b) Adsorption and kinetic studies of *Adhatoda vasica* natural dye onto woolen yarn with evaluations of Colorimetric and Fluorescence Characteristics. J Env Chem Eng 4:1780–1796

Rather LJ, Shahid I, Mohammad F (2015) Study on the application of *Acacia nilotica* natural dye to wool using fluorescence and FT-IR spectroscopy. Fibers Polym 16:1497–1505

Rather LJ, Shahid ul I, Shabbir M, Bukhari MN, Shahid M, Khan MA, Mohammad F (2016c) Ecological dyeing of woolen yarn with *Adhatoda vasica* natural dye in the presence of biomordants as an alternative copartner to metal mordants. J Environm Chem Eng 4:3041–3049

Rattanaphani S, Chairat M, Bremner JB, Rattanaphani V (2007) An adsorption and thermodynamic study of lac dyeing on cotton pretreated with chitosan. Dyes Pigm 72:88–96

Rehman F-U, Adeel S, Qaiser S, Ahmad Bhatti I, Shahid M, Zuber M (2012) Dyeing behaviour of gamma irradiated cotton fabric using Lawson dye extracted from henna leaves (*Lawsonia inermis*). Radiat Phys Chem 81:1752–1756

Samantaa AK, Agarwal P (2009) Application of natural dyes on textiles. Indian J Fibre Text Res 34:384–399

Shahid M, Mohammad F, Chen G, Tang R-C, Xing T (2016) Enzymatic processing of natural fibres: white biotechnology for sustainable development. Green Chem 18:2256–2281

Shahid M, Shahid ul I, Mohammad F (2013) Recent advancements in natural dye applications: a review. J Clean Prod 53:310–331

Shahmoradi Ghaheh F, Mortazavi SM, Alihosseini F, Fassihi A, Shams Nateri A, Abedi D (2014) Assessment of antibacterial activity of wool fabrics dyed with natural dyes. J Clean Prod 72:139–145

Sivakumar V, Anna JL, Vijayeeswarri J, Swaminathan G (2009) Ultrasound assisted enhancement in natural dye extraction from beetroot for industrial applications and natural dyeing of leather. Ultrason Sonochem 16:782–789

Tsatsaroni E, Liakopoulou-Kyriakides M, Eleftheriadis I (1998a) Comparative study of dyeing properties of two yellow natural pigments—effect of enzymes and proteins. Dyes Pigm 37:307–315

Tsatsaroni E, Liakopoulou-Kyriakides M, Eleftheriadis I (1998b) Comparative study of dyeing properties of two yellow natural pigments—effect of enzymes and proteins. Dye Pigm 37:307–315

Tutak M, Benli H (2011) Colour and fastness of fabrics dyed with walnut (*Juglans regia* L.) base natural dyes. Asian J Chem 23:566–568

Tutak M, Korkmaz NE (2012) Environmentally friendly natural dyeing of organic cotton. J Nat Fib. 9:51–59

Van Stralen T (1993) Indigo, madder & marigold: a portfolio of colors from natural dyes. Interweave Press, Colorado, USA

Vankar PS, Shanker R (2008) Ecofriendly ultrasonic natural dyeing of cotton fabric with enzyme pretreatments. Desalination 230:62–69

Vankar PS, Shanker R, Verma A (2007) Enzymatic natural dyeing of cotton and silk fabrics without metal mordants. J Clean Prod 15:1441–1450

Vankar PS, Shukla D (2012) Biosynthesis of silver nanoparticles using lemon leaves extract and its application for antimicrobial finish on fabric. Appl Nanosci 2:163–168

Vankar PS, Tiwari V, Srivastava J (2006) Extracts of stem bark of Eucalyptus globulus as food dye with high antioxidant properties. Electron J Environ Agric Food Chem 5:1664–1669

Vankar PS, Shanker R, Mahanta D, and Tiwari SC (2008) Ecofriendly sonicator dyeing of cotton with Rubia cordifolia Linn. using biomordant. Dyes Pigm. 76:207–212

Vera de Rosso V, Mercadante AZ (2009) Dyes in South America. In: Bechtold T, Mussak R (eds) Handbook of natural colorants. John Wiley & Sons, Ltd, Chichester, UK., pp 53–64

Yaman N, Özdoğan E, Seventekin N, Ayhan H (2009) Plasma treatment of polypropylene fabric for improved dyeability with soluble textile dyestuff. Appl Surf Sci 255:6764–6770

Yusuf M, Ahmad A, Shahid M, Khan MI, Khan SA, Manzoor N, Mohammad F (2012) Assessment of colorimetric, antibacterial and antifungal properties of woollen yarn dyed with the extract of the leaves of henna (*Lawsonia inermis*). J Clean Prod 27:42–50

Yusuf M, Islam S, Khan MA, Mohammad F (2016a) Investigations of the colourimetric and fastness properties of wool dyed with colorants extracted from Indian madder using reflectance spectroscopy. Optik—Intern J Light Electron Optics 127:6087–6093

Yusuf M, Khan SA, Shabbir M, Mohammad F (2016b) Developing a shade range on wool by madder (*Rubia cordifolia*) root extract with gallnut (Quercus infectoria) as Biomordant. J Nat Fiber, 1–11

Yusuf M, Mohammad F, Shabbir M, Khan MA (2017) Eco-dyeing of wool with Rubia cordifolia. Text Cloth Sustain 2:1–9

Yusuf M, Shahid M, Khan MI, Khan SA, Khan MA, Mohammad F (2015) Dyeing studies with henna and madder: a research on effect of tin (II) chloride mordant. J Saudi Chem Soc 19:64–72

Yusuf M, Shahid M, Khan SA, Khan MI, Islam S-U, Mohammad F, Khan MA (2013) Eco-dyeing of wool using aqueous extract of the roots of indian madder (*Rubia cordifolia*) as natural dye. J Nat Fibres 10:14–28

Yusuf M, Shahid ul I, Khan MA, Mohammad F (2016c) Investigations of the colourimetric and fastness properties of wool dyed with colorants extracted from Indian madder using reflectance spectroscopy. Optik—Int J Light Electron Optics 127:6087–6093

Zhang B, Wang L, Luo L, King MW (2014a) Natural dye extracted from Chinese gall—the application of color and antibacterial activity to wool fabric. J Clean Prod 80:204–210

Zhang B, Wang L, Luo L, King MW (2014b) Natural dye extracted from Chinese gall—the application of color and antibacterial activity to wool fabric. J Clean Prod 80:204–210

Zhao Q, Feng H, Wang L (2014) Dyeing properties and color fastness of cellulase-treated flax fabric with extractives from chestnut shell. J Clean Prod 80:197–203

Zheng GH, Fu HB, Liu GP (2011) Application of rare earth as mordant for the dyeing of ramie fabrics with natural dyes. Korean J Chem Eng 28:2148–2155

Zhou C-E, Kan C-W (2014) Plasma-assisted regenerable chitosan antimicrobial finishing for cotton. Cellulose 21:2951–2962

Zhou Y, Zhang J, Tang R-C, Zhang J (2014) Simultaneous dyeing and functionalization of silk with three natural yellow dyes. Ind Crop Prod

Call for Environmental Impact Assessment of Bio-based Dyeing—An Overview

Tove Agnhage and Anne Perwuelz

Abstract Bio-based dyes for textile dyeing have been widely studied on account of their eco-friendly approach. However, the use of bio-sources should not be the only parameter considered for a more environmentally sound dyeing concept. By utilizing the life cycle assessment (LCA) methodology the environmental impacts of the dyeing process can be analyzed, so as to substantiate environmental claims, but studies on this topic are generally absent. As a consequence, the eco-friendliness may be misleading if an inappropriate unit process, such as a highly resource demanding dye extraction method is used. Here, we provide an overview of key environmental issues regarding bio-based dyeing, and introduce the LCA methodology pertinent to performing environmental impact assessment of the dyeing process. We argue that each phase, from acquisition of dyestuff raw material to dyed fabric, must fit sustainability parameters by being as efficient as possible without generating waste. It is clear that assessing the environmental impacts of textile bio-based dyeing is hampered by a lack of reference studies and limited data availability. Thus, research priorities should be set to close data gaps through interdisciplinary collaboration, so as to build creditable inventories in the future.

Keywords Natural dyes · Bio-based textile dyeing · Life cycle assessment · Environmental sustainability

T. Agnhage (✉)
College of Textile and Clothing Engineering, Soochow University,
Suzhou 215006, Jiangsu, China
e-mail: tove.agnhage@ensait.fr

T. Agnhage · A. Perwuelz
ENSAIT, GEMTEX, 59056 Roubaix, France

T. Agnhage
Swedish School of Textiles, University of Borås, 501 90 Borås, Sweden

© Springer Nature Singapore Pte Ltd. 2018
S.S. Muthu (ed.), *Detox Fashion*, Textile Science and Clothing Technology,
DOI 10.1007/978-981-10-4876-0_4

1 Introduction

Life cycle assessment (LCA) has become an important tool for evaluating and communicating the environmental impacts of textiles. Among the several life cycle stages that a textile product passes through, LCA has helped to show that the largest contributions to impacts on the environment arise from the manufacturing and the use phases (Chapman 2010).

LCAs have furthermore shown that the dyeing unit process is very important with respect to impacts on the environment. For instance, a LCA study of the production of dyed cotton yarn revealed that the dyeing phase was a hotspot due to the intensive use of chemicals and energy (Bevilacqua et al. 2014). Also from Allwood et al. (2006), major environmental impacts of the textile sector arise from the use of chemicals and energy. This makes conventional dyeing an eco-issue of concern as it indeed consumes chemicals and thermal/electrical energy in addition to a great amount of water, the last around 100 times the fabric weight (Wong 2016).

Parisi et al. (2015) performed a LCA to evaluate the environmental impacts associated with a new dyeing process in comparison to a classical dyeing process. Other LCA studies deal with spin-dyeing versus conventional dyeing (Terinte et al. 2014) as well as pad-dyeing technology (Yuan et al. 2013). These studies have one thing in common: via LCA, improvement options have been proposed so as to minimize the environmental impacts of textile dyeing.

In order to improve the environmental performance of textile dyeing, researchers propose the use of bio-sourced dyes (Shahid et al. 2013) but very few include LCA (Agnhage et al. 2017). Neither do previous papers share details on important issues regarding the application of the LCA tool on bio-based dyeing.

In order to establish LCA as an environmental analysis tool, for use in textile dyeing with bio-based dyes, it is important to clarify environmental issues in the dyeing process with the perspective of LCA. So far, such an overview has not been presented. This chapter aims to close this gap.

The chapter is organized as follows: Sec. 2 introduces the LCA methodology. Section 3 provides an overview of the environmental aspects of bio-based dyeing. In Sect. 4 the importance of and difficulties regarding the use of the LCA tool on bio-based dyeing are discussed, providing an outlook for further development. Conclusions are presented in Sect. 5.

2 Life Cycle Assessment

This part presents an introduction to the LCA methodology. For further description of the LCA methodology in general, the reader is referred to ISO 1404X and the handbook produced by the European Commission (2010).

2.1 Holistic Approach

The LCA methodology holistically evaluates the environmental impacts of a product by quantifying the energy and materials used (inputs), the wastes and emissions released to environment (outputs) and the environmental impacts of those inputs and outputs over the entire life cycle (Jiménez-González et al. 2000), see Fig. 1.

The holistic approach includes not only taking into account every life cycle stage, from cradle-to-grave, it also considers multiple environmental issues (see Fig. 2) determined by the selection of impact categories. Common impact categories are global warming potential, eutrophication, acidification and photochemical oxidation (smog) among others (Ilcd 2010).

The broad scope of LCA helps to avoid a narrow view of environmental concerns and reduces the risk for burden shifting. Burden shifting; namely, shifting the environmental problem from one life cycle phase to another or one environmental issue to another (ADEME 2005; Roos et al. 2015), is illustrated in Fig. 3. From this, LCA includes a look at eco-sustainability that goes beyond an aggregation of important environmental issues. It also reflects their interlinkages and dynamics, which is different from other environmental measuring tools such as the carbon footprint (Weidema et al. 2008).

However, depending on the goal and scope of the study, life cycle stages can be excluded. The methodology can then be used to perform environmental impact assessment of one or several steps, such as cradle-to-gate or gate-to-gate, still considering multiple environmental issues.

Fig. 1 Life cycle of a (textile) product

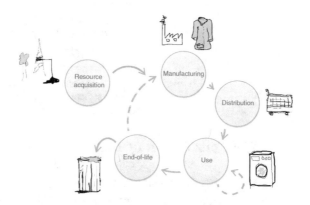

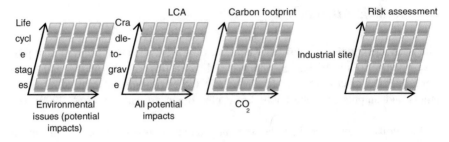

Fig. 2 Holistic approach of LCA (Roux, Irstea 2010)

Fig. 3 Diagram illustrating
burden shifting (ADEME)

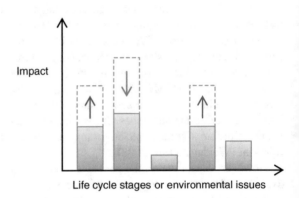

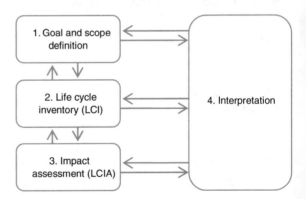

Fig. 4 The four phases of a
LCA and their interrelations
in the LCA framework (ISO
14040)

2.2 The Four Phases of LCA

According to ISO 14040 and ISO 14044 standards, a LCA comprises of four phases
which are interdependent: (1) goal and scope definition, (2) life cycle inventory
analysis (LCI), (3) life cycle impact assessment (LCIA) and, (4) interpretation (ISO
2006a, b). The relationship between these phases is illustrated in Fig. 4.

2.2.1 Phase 1: Goal and Scope

First the goal and scope must be defined. This step includes definition of the objectives, the functional unit and hypotheses. The definition of the objectives includes specifying whether the LCA tool is used for comparison or for design and optimization of a single process or product. The functional unit serves as a central element of LCA. It quantifies the function of the system studied and acts as a reference unit, by answering questions such as "what", "how much", "how well" and "how long" (Ilcd 2010). In a comparative LCA, the functional units must have the same functional performance. Otherwise a meaningful and valid comparison is not possible. However, it is not always easy to identify *the* function of the system. Systems can be multifunctional and differentiation between primary and the secondary functions may be indiscernible (Judl et al. 2012; Margni 2015).

The LCA methodology has a subjective component in several aspects, such as the goal defining and scoping as well as setting the hypotheses (Pieragostini et al. 2012). As a result, in order to get a more complete understanding of the environmental impacts of the system under study, it may be relevant to use different functional units so as to reflect various viewpoints (Cerutti et al. 2013).

The first phase also includes definition of the system boundaries. The system boundaries specify the included unit processes in the study and should include all processes required to fulfill the function.

2.2.2 Phase 2: Inventory of Resources, Emissions and Wastes

The second phase includes data collection and quantification of inputs and outputs for each unit process within the system boundaries. The resources extracted from the environment as well as emissions and wastes released into the environment, along the entire life cycle, are then grouped in an inventory. Both the quality of the primary (specific) and secondary (generic) data used to build the inventory influence the validity of the study. Weidema (1993) addresses the importance of an interdisciplinary approach (technical experts, market experts, economists etc.) when building a product's inventory. A purely technical approach may give misleading results and not reflect the actual consequence of implementing the results of the study.

2.2.3 Phase 3: Life Cycle Impact Assessment (LCIA)

In the third phase the inventory results, indicators of environmental interventions, are translated into environmental impacts. In practice, LCIA includes: choice of LCIA method (selection of relevant impact categories, category indicators, characterization factors and classification method), classification (assignment of LCI results to the relevant impact categories) and characterization (calculation of category indicator results). The characterization is calculated according to Eq. (1):

$$IR_c = \sum_s CF_{cs} * m_s \qquad\qquad ((1))$$

where IRc is the category indicator result (ex. 40 kg CO_2 eq.) for impact category
c (ex. global warming potential), CF_{cs} is the characterization factor (ex. 25 kg CO_2
eq./kg methane) that connects intervention s with impact category c and m_s is the
size of the intervention (ex. the mass of methane emitted) related to the functional
unit (Muthu 2014).

The LCIA results can be expressed either close to the environmental intervention
at the midpoint level or further away at the endpoint level. At the midpoint level,
known as the classical impact assessment method, the results are typically
expressed as equivalent values for every impact category (e.g. kg CO_2 eq. for global
warming potential). At the endpoint level, the impact categories are translated into
damage categories such as human health, resource depletion and ecosystem quality
(UNEP/SETAC 2011) and the results are expressed as damage values (e.g. DALYs
for human health). Endpoint level indicators are considered to increase in relevance,
but at the same time the uncertainty of the results increases (Bulle 2015). This is
because in general an indicator defined closer to the environmental intervention will
result in more certain modeling while an indicator more far away will provide more
relevant information linked to society's concern and areas of protection
(Georgakellos 2016; Sonnemann et al. 2004).

2.2.4 Phase 4: Interpretation

The interpretation phase is the forth and the last phase. This phase should generate a
set of conclusions, recommendations and raise significant environmental issues.

2.3 Iterative Approach

LCA is an iterative process, which allows for adjustments as a result of new
insights. The iterative character is described by the arrows back and forth between
the phases in Fig. 4. The use of LCA and its iterative approach in research will be
addressed in the next subchapter.

2.4 LCA at Research Stage

Life cycle assessment has traditionally been performed on existing large-scale
processes rather than during research and development. However, potential for
environmental improvements exists through the use of LCA at the research and
development stage. Fleischer and Schmidt (1997) presented the application of

iterative screening using the LCA methodology in an eco-design tool, so as to include environmental aspects during product development. More recently, the use of a life cycle oriented approach facilitated the uptake of innovative technologies and enabled value chain upgrading in a textile dyeing industry (Angelis-Dimakis et al. 2016). The contribution of LCA to product innovation and value creation opportunities, such as lower environmental impacts, has also been highlighted in Manda et al. (2015). Furthermore, Foulet et al. (2015) and Pasquet et al. (2014) have shown that LCA is a useful tool to assess process impacts at the early research stage. The great potential of using LCA in early research has also been reported in Zerazion et al. (2016). Their work (Zerazion et al. 2016) is particularly noteworthy in the context of the present overview. This is because it considered the production, at lab scale, of the natural dye *Curcuma longa L.* and shows how LCA can help in establishing a more eco-friendly production of a target compound from natural resources.

Nevertheless, the use of LCA within research, in particular within early research, faces several challenges as addressed hereafter.

Hetherington et al. (2014) highlight four main issues in using LCA for early research: (1) comparability, (2) scale, (3) data and, (4) uncertainty. The comparability issue includes for example the challenge that the new material may not be functionally equivalent to that which it replaces. The scale issue includes the challenge of laboratory production often being completed as a batch process with high impacts on energy consumption for startup and shutdown. The data and uncertainty challenges consist of the use of secondary data due to primary data taking too long to gather, or lack of data for new materials. Lack of data is a familiar situation for LCA practitioners, but for novel processes wider data gaps that contribute to the level of uncertainty may be expected.

Because of this, simplified approaches are essential in order to perform LCA for early research (Fleischer and Schmidt 1997). Consequently, it becomes paramount to communicate uncertainties so as to let others better judge the results.

2.5 Uncertainty and Data Quality

LCA tries to model the reality and, through its iterative character, aims first to be accurate (screening) and works then on precision (detail LCA). It is preferable to be imprecisely accurate than precisely inaccurate (Humbert et al. 2015). In order to estimate results, such as evaluating how well the LCA model reflects the reality or the actual consequence of implementing the results of the investigation, an uncertainty analysis should be performed.

If the LCA database does not include results on uncertainty, another way is to use a pedigree matrix with data quality indicators. Weidema and Wesnaes (1996) have introduced five indicators to describe the data quality: reliability, completeness, temporal, geographical and technological. Details on how to apply the data quality matrix can be found in Weidema and Wesnaes (1996), and in Weidema (1998).

Noteworthy, the guidelines for the product environmental footprint (PEF) implementation recommend six data quality criteria by adding methodology to the other five (PEF 2012).

It is envisaged that a multidisciplinary approach will improve the data quality and reduce uncertainties. The need for interdisciplinary collaboration in LCAs of bio-based dyeing will be discussed in Sect. 4. Before that, the next part (Sect. 3) will present environmental aspects of bio-based dyeing.

3 Environmental Aspects of Bio-based Dyeing

This part presents an overview of bio-based dyeing of textiles. In particular, it will address the use of coloring species from plants and environmental issues it raises. The subchapter however, starts with a brief introduction to the classification of bio-sourced dyes.

3.1 Classification and Context

Natural dyes can be classified in a number of ways, such as on the basis of hue or according to chemical constitution. Yellow is probably the most abundant dyestuff hue found in nature and more common than red or blue ones. The yellow dyes, which can be found in chamomile or weld for example, are mainly flavonoids (Peters et al. 2015) whereas red dyes, such as the dyes extracted from madder root, are almost invariably based on anthraquinones and its derivatives (Bechtold 2009). Blue dyes are commonly indigoid dyes, such as indigo.

Classification can also be done based on the method of application; such as, mordant dyes and vat dyes. Moreover, the source of origin may be used for natural dye classification. The source of origin considers plants, animals, minerals or microbes.

Most commonly bio-based dyes originate from plants (Saxena and Raja 2014) and plant-based dyes are the once addressed in this work. Moreover, based on Bechtold et al. (2007), the present chapter covers the use of both primary plant sources and byproducts. A schematic representation of how they are used for textile dyeing is illustrated in Fig. 5. Environmental aspects related to the three phases: raw material acquisition (cultivation or collection), dye extraction and the dyeing procedure are addressed henceforth.

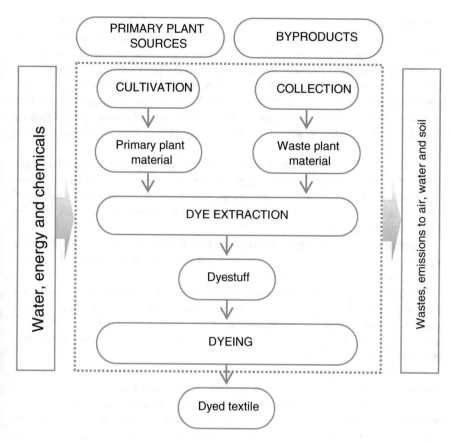

Fig. 5 Schematic representation of textile dyeing using plant-based dyes

3.2 Raw Material Acquisition (Cultivation/Collection)

3.2.1 Primary Plant Sources

Primary plant sources are tinctorial crops cultivated as colorants. The coloring species offered by these crops can be obtained from different parts of the plant for example, the roots, flowers and leaves.

The use of material from plants may contribute to sustainable development as fossil resources are conserved (Ganglberger 2009). Nevertheless, promotion of plant-based dyes without ensuring increased availability of the dye-bearing material can lead to overexploitation. Moreover, the environmental impacts of water and land use should be considered. Often those impacts are dealt with in a simplified manner (Schmidt et al. 2015). Sandin et al. (2013) assessed the water and land use impacts of bio-based textile fibers and revealed that water extracted from water stressed environments will give higher impacts. Furthermore, biodiversity impacts may be higher from transformation than occupation of land. Biogenic carbon storage is another important issue. Based on Pawelzik et al. (2013), carbon storage in bio-based materials remains controversial, taking account of it is recommended.

The environmental impacts are also related to the inputs of chemicals, such as fertilizers as soil nutrients and chemicals for weed control. Those chemicals are responsible for soil and water pollution. The LCA tool can help to find adequate solutions for trade-offs between chemical use and crop yield; for example, applying fertilizers may benefit the plant growth, but may also increase the environmental impacts.

Some well-screened natural dye crops, for the geographical scope of Europe, which can be used for industrial applications are woad, ai, madder and weld (Biertümpfel and Wurl 2009). One reason for their cultivation is the amount of coloring species in the plant. The bio-dye amount should be high enough for environmental benefits.

Moreover, to improve the environmental profile of the dye, the plant wastes should be utilized (e.g. as animal feed) and the environmental impacts allocated to the different products from the plant.

The process of handling the production and use of primary plant sources for dyeing deserves further attention, in order to be as efficient as possible without generating waste. Hence, a recommendation for future work would be to focus on new perspectives of growing tinctorial plants and to explore moves toward permaculture or hydroculture.

Permaculture emphasizes the sustainable use of natural resources and the practice of working *with* rather than *against* nature (Akhtar et al. 2016). Here it is relevant to address a work which, in pursuing the sustainable use of curcumin from the turmeric plant, studied the application of chitosan. Chitosan protected the plant and enhanced the growth and curcumin yield, and was labeled as a promising eco-friendly compound for curcumin production (Anusuya and Sathiyabama 2016).

Hydroculture (soilless culture system) may be an option to improve the environmental profile, as this cultivation condition enables precise control of nutrient elements. Although it has been studied for plants relating to the food-industry, there are limited studies available on hydroculture for the production of dye-bearing plants for textile dyeing (Asghari 2015).

3.2.2 Byproduct Plant Sources

The sustainability concept of zero emission and zero waste; namely, that one industry can consume another industry's wastes, has encouraged research on bio-based residues as coloring matter for textiles. Some have explored coloring-bearing materials from the food and beverage industries, such as olive oil wastewater (Haddar et al. 2014a), wine waste (Baaka et al. 2015), peels (Yi and Yoo 2010), date pits powder (Guesmi et al. 2016) and red pepper by products (Ksibi et al. 2015). Conversely, others have studied the use of byproducts from the wood industry (Bechtold et al. 2007; Baliarsingh et al. 2013; Karaboyaci 2014).

Indeed, the use of waste materials has attracted attention and today one can find commercial dyes available based for example on almond husk (http://textiles. archroma.com).

Efficient use of byproducts represents an opportunity for the eco-friendlier production of natural dyes as it minimizes waste production, reduces emissions associated with dye production and supports the transition towards a circular economy in textiles. However, the colors provided through byproducts may differ from the colors obtained with primary sources. In order to have a complete color palette, both residual plant materials and primary plant materials may be used (Shahid-ul-Islam et al. 2013).

3.3 Dye Extraction

Tinctorial plants contain coloring species but also other non-coloring plant constituents, starch among others. The non-coloring matter reduces the quality of the dye. Hence, separating the coloring species from the non-coloring ones through dye extraction is essential, in order to obtain dyes with high quality.

One of the more critical environmental considerations here is the selection of extraction solvent and the question regarding dye yield versus environmental impacts. Mussak and Bechtold (2009) suggest explicit water extraction, because the input of chemicals or organic solvents may lead to contaminated wastewater and require complicated wastewater treatment. However, water as the solvent limits the amount of extracted dyestuff to water-soluble compounds. If organic solvents are chosen (such methanol or ethanol) potentially an increased dyestuff yield can be obtained (Ganglberger 2009). Acid or alkali can also be added to enhance and enrich the dye yield. Nevertheless, the use of organic solvents contribute to the greenhouse effect, toxicity and water pollution. Furthermore, a considerable amount of solvent may remain in the waste, which eliminates the possibility to utilize the waste as animal feed (Saxena and Raja 2014). So, on the one hand the use of organic solvent and chemicals may enable higher dye yield but, on the other hand, it involves the presence of solvent or chemical residuals and their potential environmental impacts.

To minimize the amount of solvent and the environmental impacts of effluents, the solvent is usually recycled. However, those recycling processes need energy for solvent regeneration (from distillation or filtration techniques). Consequently, there is a question regarding energy use versus effluent pollution and raw material depletion.

Solvent recycling and good selectivity are possible using the conventional Soxhlet extraction technique. One of the main disadvantages of using Soxhlet however, is the high consumption of energy because of long extraction time with high temperature (Shahid-ul-Islam et al. 2013), even under optimized process conditions (Sinha et al. 2012). A description of the Soxhlet technique can be found in the review on extraction techniques presented in Shahid et al. (2015).

In order to minimize the environmental impacts from dye extraction, unconventional extraction methods have been researched. Shahid-ul-Islam et al. (2013) address ultrasonic, microwave and supercritical carbon dioxide extraction as three promising methods since these prove to deliver better results than conventional extraction while using less resources. Results from literature reveal improvements such as reduced extraction time from 60 min (aqueous extraction at boil) to 5–15 min (ultrasound) (Sheikh et al. 2016). Sivakumar et al. (2011) report a 13–100% increased extraction efficiency of dyes from different plant materials, due to the use of ultrasound instead of magnetic stirbar. A comparison between classical reflux extraction and novel ultrasound extraction has been described in Cuoco et al. (2009), and their results pointed out higher dye yield and shorter extraction time when replacing the classical method with the novel one.

The interest in novel techniques, for the extraction of natural compounds, can be further confirmed through the recently presented review paper on advances for the extraction of plant-based anthraquinones (Duval et al. 2016) as well as phenolics (Shahid et al. 2015).

Natural dyes are not a single chemical entity but a mixture of closely related compounds, which can vary in their relative content according to the age of the plant or climate conditions. Indeed, the dye constitution of natural sources is not fixed. Thus, to ensure high purity and quality of the extracted dyes, identification and characterization of the dyes are important.

High-performance chromatography (HPLC) technique helps in understanding the chemical composition of natural dyes and has been used for the characterization of anthraquinone glycosides and aglycones in extracts of madder roots (Derksen et al. 2004, 1998) as well as for the development of an industrial usable method to quickly assess the flavone content of weld samples (Villela et al. 2011).

3.4 Dyeing Procedure

This part looks at three factors influencing the environmental profile of the dyeing procedure: the purity of dyestuff, the dyeability of the textile substrate and the dyeing method. The section also addresses durability/quality aspects.

3.4.1 Purity of Dyestuff

Looking at the intensive use of water, energy and chemicals in the dyeing procedure, minimum resource consumption can only be achieved through employing dyestuff with high purity. Natural impurities may lead to incorrect dyeing (for instance uneven dyeing or shade to shade variation) that requires corrections by shading or, in the worst case, overdyeing. Overdyeing will lead to additional impacts on the environment because the repeated use of water, energy and chemicals, and may offset the initial potential environmental benefits (Youssef 2016). It is thus, important that the dye extraction provides high quality dyes.

3.4.2 Dyeability of Textile Substrate

Traditionally, the dyeing process with natural dyes involves the use of mordants, a name derived from Latin mordere, meaning 'to bite' (Singh and Bharati 2014). These chemicals, such as metal salts, form a complex with the dye, which improves the dye uptake onto the textile fiber.

The importance of mordants has been in improving the dyeing process, with respect to durability/fastness properties and color gamut (Shahid et al. 2013), but metal salts released into wastewater is an unwanted consequence because it may cause water pollution (GOTS). To reduce the risk for pollution, textile surface modification and the use of bio-mordants have been studied as alternative methods to metal mordanting.

Haddar et al. (2014b) studied natural dyeing of surface modified cotton by cationization. Their work evaluated the environmental performance of the dyeing bath by measuring the polyphenol concentration, chemical oxygen demand (COD) and biological oxygen demand (BOD). It was concluded that the developed dyeing process, using cationized cotton and response surface methodology, reduced environmental loads. Another study dealing with surface modification is the one in Khan et al. (2014). Their work (Khan et al. 2014) presented reduced amount of mordants in the natural dyeing of cotton through the use of gamma ray treatment of the textile fabric as well as of the dye. High-energy irradiation methods (gamma ray, ultrasound and plasma among others) have attracted attention in literature. Recent advances in their use, for inducing surface modification of textiles, can be found in Shahid-ul-Islam and Mohammad (2015).

Instead of utilizing irradiation-based methods for modification of the textile surface, Mehrparvar et al. (2016) replaced the use of metal mordants in wool dyeing with a dendrimer-modified chitosan (CS-PPI) treatment of the textile. The CS-PPI thus acted as a biomordant.

Biomordants have been addressed in several papers so as to sustain the environmentally friendly approach of natural dyeing. Rather et al. (2016) present gallnut as an interesting alternative to metal salts for dyeing wool. Moreover, in order to dye wool, rosemary leaves and other biomordants have been studied together with almond shell extracts (İşmal et al. 2014). The biomordants were expressed as a promising concept, (İşmal et al. 2014), though from light fastness evaluation, the color fastness to light was low. Contrary to others (previously cited), Tchinda et al. (2014) studied dyeing (using red padouk extract from wood industrial waste) without any mordant, arguing that the tannins in the padouk could act as a mordant.

3.4.3 Dyeing Method

Conventional dyeing with mordants can be applied as a pre-, post- or meta-procedure. The two first involve mordanting as an additional step, making the dyeing procedure a two-step process. Instead meta-mordanting is a one-bath method in which the mordant is applied directly into the dyeing bath. In Ganglberger (2009), meta-mordanting is preferred from an ecological point of view, since less water and energy is required. However, although the one-bath option seems to be favorable, one disadvantage according to Mussak and Bechtold (2009) is the difficulty in re-using the dyeing bath for a second dyeing because it is contaminated with mordants.

Some papers deal with improved natural dyeing, in terms of reduced use of water and energy. Mansour and Heffernan (2011) used ultrasonic-energy to assist the dyeingbath. This reduced dyeing time and temperature while at the same time improving the exhaustion and fixation of natural dye onto silk. Barani and Maleki (2011) studied low temperature plasma and surface-active biological lipids (liposomes) as an environmentally friendly treatment to improve the dyeability of wool with natural dye. The use of liposomes in the natural dyeing of wool has also been studied in Montazer et al. (2007). Their work (Montazer et al. 2007) showed that liposomes reduced the dyeing temperature and, under a certain concentration, improved the color strength of the dyed fabric. Moreover, waterless dyeing using supercritical fluids (SC-CO_2) as the solvent for plant-based dyestuff (Guzel and Akgerman 2000) and reuse of wastewater from natural dyeing (Shams-Nateri 2011) have been reported as promising alternatives to conventional dyeing methods.

3.4.4 Color Durability and Quality

From the studies presented above (Sects. 3.4.2 and 3.4.3), there is an attempt to improve the material and energy efficiency in the textile sector dealing with bio-sourced dyes. However, as addressed in environmental management (ISO/TR 2002), in pursuing a more sustainable development, it is necessary to consider not only conservation of resources but also the functionality of the product and its durability properties. Increasing the durability/quality can reduce environmental impacts as it may delay the disposal of the product and the arrival of another one using water, energy and chemicals (Muthu et al. 2013). In order to not have a larger than necessary impact on the environment, one should emphasize the balance between environmental impacts and quality.

4 Final Discussions

This part will highlight the current challenges in applying LCA as a tool for the eco-sustainable use of plant-based dyes in textile dyeing. What is the importance of using LCA and what are the difficulties?

4.1 The Importance—Value Creation Using LCA

Textile dyeing refers to a unit process in the multi-step production chain of textiles, which has usually been acknowledged for adding value to the end product in terms of aesthetics. A current challenge facing the textile sector is however, to obtain dyed textiles without compromising on environmental sustainability. The LCA tool can help to find an optimum compromise between the positive aesthetic value and the negative environmental externality, and add value in terms of reduced environmental impacts. From this, it is envisaged that the LCA tool will be of increased importance in the future.

From a LCA point of view, the environmental impacts of bio-based dyeing are not only an inherent consequence of the chemistry in the dyeing house. The environmental performance depends on several issues related to the acquisition of plant material, dye extraction, dyeing method and finally the quality. It is important to include all these steps in the system boundaries. The LCA tool can then help to prevent pollution transfer from one life cycle phase to another. For example, the LCA tool may ensure that a reduced impact in agriculture is not at the expense of worsening the extraction phase due to a lower dye yield in the plant, or that a reduced impact in dye extraction is not at the expense of worsening the dyeing phase due to low quality dyes. Furthermore, the tool can help to answer questions and resolve trade-offs within each phase for example, support decision making regarding whether or not to use solvent recycling in the dye extraction phase.

All things considered, the importance of applying environmental impact assessment lies in value creation. A value creation based on reducing the environmental impacts in each life cycle phase from plant material acquisition to dyed fabric, while at the same time preventing pollution transfer between the stages.

4.2 The Difficulties—Fill Data Gaps

In order to carry out LCA of bio-based dyed textiles, before collecting inventory data, the functional unit has to be defined. When considering textiles dyed with bio-dyes, there are many properties that can be attributed: color, shade, uniformity and durability but also other functional properties such as antibacterial activity and ultraviolet protection ability (Shahid et al. 2013). Because of the multifunctional performance it may not be obvious how to distinguish the primary function from secondary ones, and this difficulty should be noted.

Once the FU has been defined, one can start data collection and the quantification of inputs and outputs for each unit process included in the system boundaries. The unit processes from acquisition of plant material to dyed fabric will be addressed hereafter, from the viewpoint of building the LCA inventory.

Eco-sustainability issues in the cultivation and acquisition of tinctorial plant-material are land use, water consumption, soil pollution and waste generation. It is important to remember that chemicals may improve the crop yield while worsening the pollution and the by products generated should be valorized and allocations considered. New research is needed to optimize the cultivation process so as to answer questions concerning land use, efficiency and yield, as well what to do with the wastes that originate. Contributions from agro-researchers, among others, are essential here, as it can help to build a creditable inventory.

Regarding the dye extraction phase, two questions have been raised. One regards the use of organic solvents and the potentially increased dyestuff yield but at the same time increased global warming potential and toxicity. The other question considers recycling of the solvent, which is necessary in order to reduce the environmental burdens from the effluents. However, the recycling is energy consuming. New resource efficient extraction techniques should be developed by chemists and input and output data shared to the LCI.

The bio-dyeing phase of textiles needs water and energy, which can be reduced through optimized dyeing duration, temperature, or with new processes. Most of these need to be studied through LCA to have a relevant overview. However, the color of the dyed fabric and its durability may vary depending on the cultivation conditions of the dyes. Their development will depend on consumer acceptance and thus, skills from the sociology and fashion research arena would improve the inventory.

From this, an interdisciplinary collaboration is called for so as to build a successful LCA model and inventory, in terms of reflecting, to the largest extent possible, the actual consequences of implementing the results. Firstly then value can be created, value as in improved eco-sustainability of bio-based dyeing of textiles.

5 Conclusions

This overview paper calls for the introduction of the life cycle assessment (LCA) tool to bio-based dyeing of textiles, in order to encourage the use of bio-sourced dyes in a more sustainable way. Environmental aspects of the dyeing process have been addressed, from the perspective of LCA, including the importance of LCA as well as the difficulties that may arise.

The importance, of using the LCA tool lies in value creation, in terms of reducing environmental impacts. The textile sector is to some extent familiar with the use of LCA and how, through its holistic approach, it can help to reduce impact on the environment. Nevertheless, the tool has yet not reached the research field dealing with bio-based dyeing of textiles.

It is however, obvious that LCA at the research stage of textile dyeing with bio-based coloring species brings challenges. One of the more difficult issues is to build a representative inventory, due to the lack of data. Research priorities should therefore, be set to close data gaps through interdisciplinary collaboration, so as to build creditable inventories in the future.

Acknowledgements This work was realized within the framework of Sustainable Management and Design for Textiles, financed by European Erasmus Mundus program and the EU Window Chinese Government Scholarship. The authors wish to thank Nemeshwaree Behary and Christian Catel (ENSAIT-Gemtex laboratory) for their contributions.

References

ADEME (2005) Introduction de l'analyse de cycle de Vie (ACV). In: Pasquet V (2012) Contribution àl'étude de l'impact environnemental de procédés de traitements de textiles par l'outil d'analyse du cycle de vie. Dissertation, Lille University, France

Agnhage T, Perwuelz A, Behary N (2017) Towards sustainable Rubia tinctorum L. dyeing of woven fabric: how life cycle assessment can contribute. J Clean Prod 141:1221–1230. doi:10.1016/j.jclepro.2016.09.183

Akhtar F, Lodhi SA, Khan SS et al (2016) Incorporating permaculture and strategic management for sustainable ecological resource management. J Environ Manage 179:31–37. doi:10.1016/j.jenvman.2016.04.051

Allwood JM, Laursen SE, de Rodriguez CM et al (2006) Well dressed? The present and future sustainability of clothing and textiles in the United Kingdom. Institute for Manufacturing, University of Cambridge, UK

Angelis-Dimakis A, Alexandratou A, Balzarini A (2016) Value chain upgrading in a textile dyeing industry. J Clean Prod 138:237–247. doi:10.1016/j.jclepro.2016.02.137

Anusuya S, Sathiyabama M (2016) Effect of chitosan on growth, yield and curcumin content in turmeric under field condition. Biocatal Agric Biotechnol 6:102–106. doi:10.1016/j.bcab.2016.03.002

Asghari R (2015) Effects of nutritional elements level on nutritional characters and phytochemistry of strawberry in hydroculture. Ital J Food Sci 27:1–5

Baaka N, Ticha MB, Haddar W et al (2015) Extraction of natural dye from waste wine industry: optimization survey based on a central composite design method. Fiber Polym 16(1):38–45. doi:10.1007/s12221-015-0038-5

Baliarsingh S, Jena J, Das T et al (2013) Role of cationic and anionic surfactants in textile dyeing with natural dyes extracted from waste plant materials and their potential antimicrobial properties. Ind Crops Prod 50:618–624. doi:10.1016/j.indcrop.2013.08.037

Barani H, Maleki H (2011) Plasma and ultrasonic process in dyeing of wool fibers with madder in presence of lecithin. J Dispers Sci Technol 32:1191–1199. doi:10.1080/01932691.2010.505525

Bechtold T (2009) Natural colorants—quinoid, naphthaquinoid and anthraquinoid dyes. In: Bechtold M, Mussak R (eds) Handbook of natural colorants. Wiley, Hoboken, p 162

Bechtold T, Mahmud-Ali A, Mussak RAM (2007) Reuse of ash-tree (Fraxinus excelsior L.) bark as natural dyes for textile dyeing: process conditions and process stability. Color Technol 123:271–279. doi:10.1111/j.1478-4408.2007.00095.x

Bevilacqua M, Ciarapica FE, Mazzuto G et al (2014) Environmental analysis of a cotton yarn supply chain. J Clean Prod 82:154–165. doi:10.1016/j.jclepro.2014.06.082

Biertümpfel A, Wurl G (2009) Dye plants in Europe. In: Bechtold M, Mussak R (eds) Handbook of natural colorants. John Wiley & Sons Ltd, Chichester, UK, pp 39–52

Bulle C (2015) Applications of life cycle assessment to regional issues. Lecture presented at the 7th international conference on life cycle management, Bordeaux, France, 30 Aug–2 Sep 2015

Cerutti AK, Bruun S, Donno D, et al (2013) Environmental sustainability of traditional foods: the case of ancient apple cultivars in Northern Italy assessed by multifunctional LCA. J Clean Prod 52:245–252. http://dx.doi.org.lib.costello.pub.hb.se/10.1016/j.jclepro.2013.03.029

Chapman A (2010) Review of life cycle assessments of clothing. A report for Mistra Future Fashion, Oakdene Hollins Research & Consulting. File reference number: MIST01 232 LCA review.doc. http://www.oakdenehollins.co.uk/media/232/2010_mistra_review_of_life_cycle_assessments_of_clothing.pdf

Cuoco G, Mathe C, Archier P et al (2009) A multivariate study of the performance of an ultrasound-assisted madder dyes extraction and characterization by liquid chromatography-photodiode array detection. Ultrason Sonochem 16:75–82. doi:10.1016/j.ultsonch.2008.05.014

Derksen GCH, Lelyveld GP, van Beek TA et al (2004) Two validated HPLC methods for the quantification of alizarin and other anthraquinones in *Rubia tinctorum* cultivars. Phytochem Anal 15:397–406

Derksen GCH, van Beek TA, de Groot Æ et al (1998) High-performance liquid chromatographic method for the analysis of anthraquinone glycosides and aglycones in madder root (*Rubia tinctorum* L.). J Chromatogr A 816(2):277–281.http://dx.doi.org.lib.costello.pub.hb.se/10.1016/S0021-9673(98)00492-0

Duval J, Pecher V, Poujol M et al (2016) Research advances for the extraction, analysis and uses of anthraquinones: A review. Ind Crop Prod 94:812–833. doi:10.1016/j.indcrop.2016.09.056

Fleischer G, Schmidt WP (1997) Iterative screening LCA in an eco-design tool. 6th SETAC-Europe Meeting. Int J LCA 2(1):20–24

Foulet A, Birot M, Sonnemann G et al (2015) Life cycle assessment of producing emulsion-templated porous materials from Kraft black liquor-comparison of a vegetable oil and a petrochemical solvent. J Clean Prod 91:180–186. doi:10.1016/j.jclepro.2014.12.035

Ganglberger E (2009) Environmental aspects and sustainability. In: Bechtold M, Mussak R (eds) Handbook of natural colorants. Wiley, Hoboken, p 162

Georgakellos DA (2016) Life cycle analysis as a management tool in environmental systems. In: Sarkar D, Datta R, Mukherjee A et al (ed) An integrated approach to environmental management. John Wiley & Sons, Ltd., pp 441–464

Global Organic Textile Standard (GOTS) Version 5.0 (2017) Available via www.global-standard.org. Accessed 14 Mar 2017

Guesmi A, Dhahri H, Hamadi NB (2016) A new approach for studying the dyeability of a multifibers fabric with date pits powders: a specific interest to proteinic fibers. J Clean Prod 133:1–4. doi:10.1016/j.jclepro.2016.05.075

Guzel B, Akgerman A (2000) Mordant dyeing of wool by supercritical processing. J Supercrit Fluids 18:247–252

Haddar W, Baaka N, Meksi N et al (2014a) Optimization of an ecofriendly dyeing process using the wastewater of the olive oil industry as natural dyes for acrylic fibres. J Clean Prod 66:546–554. doi:10.1016/j.jclepro.2013.11.017

Haddar W, Elksibi I, Meksi N et al (2014b) Valorization of the leaves of fennel (*Foeniculum vulgare*) as natural dyes fixed on modified cotton: A dyeing process optimization based on a response surface methodology. Ind Crops Prod 52:588–596. doi:10.1016/j.indcrop.2013.11.019

Hetherington AC, Borrion AL, Griffiths OG et al (2014) Use of LCA as a development tool within early research: challenges and issues across different sectors. Int J Life Cycle Assess 19:130–143. doi:10.1007/s11367-013-0627-8

Humbert S, Fantke P, Magaud V (2015) Using the USEtox[i] model in life cycle assessment. Lecture presented at the 7th international conference on life cycle management, Bordeaux, France, 30 Aug–2 Sep 2015 http://textiles.archroma.com Accessed 19 Jan 2017

Ilcd (2010) International Reference Life Cycle Data System (ILCD) Handbook— general guide for life cycle assessment—detailed guidance, 1st ed. Publications Office of the European Union, Luxembourg. European Commission, 2010. doi:10.2788/38479

İşmal ÖE, Yıldırım L, Özdoğan E (2014) Use of almond shell extracts plus biomordants as effective textile dye. J Clean Prod 70:61–67. doi:10.1016/j.jclepro.2014.01.055

ISO (International Organization for Standardization) (2006a) ISO14040–Environmental Management–Life Cycle Assessment Principles and Framework. ISO, Geneva

ISO (International Organization for Standardization) (2006b) ISO14044–Environmental Management—Life Cycle Assessment—Requirements and Guidelines. ISO, Geneva

ISO (International Organization for Standardization) 2002(E) ISO/TR14062–Environmental Management—Integrating environmental aspects into product design and development. Prepared by Technical Committee ISO/TC 207, Environmental management

Jiménez-González C, Kim S, Overcash MR (2000) Methodology for developing gate-to-gate Life cycle inventory information. Int J Life Cycle Assess 5:153–159

Judl J, Mattila T, Seppala J et al (2012) Challenges in LCA comparisons of multifunctional electronic devices. IEEE Electron Goes Green 61(1):1–5

Khan AA, Iqbal N, Adeel S et al (2014) Extraction of natural dye from red calico leaves: Gamma ray assisted improvements in colour strength and fastness properties. Dyes Pigm 103:50–54. doi:10.1016/j.dyepig.2013.11.024

Karaboyaci M (2014) Recycling of rose wastes for use in natural plant dye and industrial applications. J Text I 105(11):1160–1166. doi:10.1080/00405000.2013.876153

Ksibi IE, Slama RB, Faidi K et al (2015) Mixture approach for optimizing the recovery of colored phenolics from red pepper (*Capsicum annum* L.) by-products as potential source of natural dye and assessment of its antimicrobial activity. Ind Crops Prod 70:34–40. doi:10.1016/j.indcrop.2015.03.017

Manda BMK, Worrell E, Patel MK (2015) Prospective life cycle assessment of an antibacterial T-shirt and supporting business decisions to create value. Resour Conserv Recy 103:47–57. http://dx.doi.org.lib.costello.pub.hb.se/10.1016/j.resconrec.2015.07.010

Mansour HF, Heffernan S (2011) Environmental aspects on dyeing silk fabric with sticta coronata lichen using ultrasonic energy and mild mordants. Clean Technol Envir 13(1):207–213. doi:10.1007/s10098-010-0296-2

Margni M (2015) Life cycle assessment for products, processes and services. Lecture presented at the 7th international conference on life cycle management, Bordeaux, France, 30 Aug–2 Sep 2015

Mehrparvar L, Safapour S, Sadeghi-Kiakhani M et al (2016) A cleaner and eco-benign process for wool dyeing with madder, Rubia tinctorum L., root natural dye. Int J Environ Sci Technol 13:2569–2578. doi:10.1007/s13762-016-1060-x

Montazer M, Taghavi FA, Toliyat T et al (2007) Optimization of dyeing of wool with madder and liposomes by central composite design. J Appl Polym Sci 106:1614–1621. doi:10.1002/app.26841

Mussak RAM, Bechtold T (2009) Natural colorants in textile dyeing. In: Bechtold M, Mussak R (eds) Handbook of natural colorants. Wiley, Hoboken, p 162

Muthu SS (2014) Assessing the environmental impact of textiles and the clothing supply chain. Woodhead Publishing Series, Textiles Number, p 157

Muthu SS, Li Y, Hu JY et al (2013) Modelling and quantification of eco-functional index: The concept and applications of eco-functional assessment. Ecol Indic 26:33–43. doi:10.1016/j.ecolind.2012.10.018

Parisi ML, Fatarella E, Spinelli D et al (2015) Environmental impact assessment of an eco-efficient production for coloured textiles. J Clean Prod 108:514–524. doi:10.1016/j.jclepro.2015.06.032

Pasquet V, Behary N, Perwuelz A (2014) Environmental impacts of chemical/ecotechnological/ biotechnological hydrophilisation of polyester fabrics. J Clean Prod 65:551–560. doi:10.1016/ j.jclepro.2013.06.052

Pawelzik P, Carus M, Hotchkiss J et al (2013) Critical aspects in the life cycle assessment (LCA) of bio-based materials—Reviewing methodologies and deriving recommendations. Resour Conserv Recy 73:211–228. doi:10.1016/j.resconrec.2013.02.006

PEF (2012) Product Environmental Footprint (PEF) Guide, Consolidated version. http://ec.europa. eu/environment/eussd/pdf/footprint/PEF%20methodology%20final%20draft.pdf

Peters CA, Sgrott RAG, Peters RR et al (2015) Production of Wilbrandia ebracteata extract standardized in flavonoids and dihydrocurcubitacin and assessment of its topical anti-inflammatory activity. Ind Crop Prod 69:123–128. doi:10.1016/j.indcrop.2015.01.018

Pieragostini C, Mussati MC, Aguirre P (2012) On process optimization considering LCA methodology. J Environ Manage 96:43–54. doi:10.1016/j.jenvman.2011.10.014

Rather LJ, Shahid-ul-Islam, Mohd Shabbir M et al (2016) Ecological dyeing of Woolen yarn with *Adhatoda vasica* natural dye in the presence of biomordants as an alternative copartner to metal mordants. JECE 4(3):3041–3049

Roos S, Posner S, Jönsson C et al (2015) Is unbleached cotton better than bleached? Exploring the limits of life-cycle assessment in the textile sector. Cloth Text Res J 33(4):231–247. doi:10. 1177/0887302X15576404

Roux, Irstea (2010) In: Sonnemann G (2015) Using LCA as a metrics in advancing and scaling up Green Chemistry research. Lecture presented at the 7th international conference on life cycle management, Bordeaux, France, 30 Aug–2 Sep 2015

Sandin G, Peters GM, Svanström M (2013) Moving down the cause-effect chain of water and land use impacts: an LCA case study of textile fibres. Resour Conserv Recycl 73:104–113. doi:10. 1016/j.resconrec.2013.01.020

Saxena S, Raja ASM (2014) Natural dyes: sources, chemistry, application and sustainability issues. In: Muthu SS (ed) Roadmap to sustainable textiles and clothing. Springer Science, Singapore, pp 37–80. doi:10.1007/978-981-287-065-0_2

Schmidt JH, Weidema BP, Brandão M (2015) A framework for modelling indirect land use changes in life cycle assessment. J Clean Prod 99:230–238. doi:10.1016/j.jclepro.2015.03.013

Shahid M, Yusuf M, Mohammad F (2015) Plant phenolics: a review on modern extraction techniques. Recent progress in medicinal plants, vol 41-analytical and processing techniques. Studium Press LLC, USA, pp 265–287

Shahid M, Shahid-ul-Islam, Mohammad F (2013) Recent advancements in natural dye applications: a review. J Clean Prod 53:310–331. doi:10.1016/j.jclepro.2013.03.031

Shahid-ul-Islam, Mohammad F (2015) High-energy radiation induced sustainable coloration and functional finishing of textile materials. Ind Eng Chem Res 54(15):3727–3745. doi:10.1021/ acs.iecr.5b00524

Shahid-ul-Islam, Shahid M, Mohammad F (2013) Perspectives for natural product based agents derived from industrial plants in textile applications–a review. J Clean Prod 57:2–18. doi:10. 1016/j.jclepro.2013.06.004

Shams-Nateri A (2011) Reusing wastewater of madder natural dye for wool dyeing. J Clean Prod 19:775–781. doi:10.1016/j.jclepro.2010.12.018

Sheikh J, Jagtap PS, Teli MD (2016) Ultrasound assisted extraction of natural dyes and natural mordants vis a vis dyeing Fiber Polym 17(5):738–743. doi:10.1007/s12221-016-5031-0

Singh HB, Bharati KA (2014) Mordants and their applications. In: Singh HB, Bharati KA (eds) Handbook of natural dyes and pigments. Woodhead Publishing, India Pvt. Ltd., pp 18–28. doi:10.1016/B978-93-80308-54-8.50004-6

Sinha K, Saha PD, Datta S (2012) Extraction of natural dye from petals of Flame of forest (*Butea monosperma*) flower: process optimization using response surface methodology (RSM). Dyes Pigm 94(2):212–216. doi:10.1016/j.dyepig.2012.01.008

Sivakumar V, Vijaeeswarri J, Lakshmi JA (2011) Effective natural dye extraction from different plant materials using ultrasound. Ind Crop Prod 33:116–122. doi:10.1016/j.indcrop.2010.09. 007

Sonnemann G, Castells F, Schuhmacher M (2004) Integrated life-cycle and risk assessment for industrial process. Lewis Publishers, Boca Raton

Terinte N, Manda BMK, Taylor J et al (2014) Environmental assessment of coloured fabrics and opportunities for value creation: spin-dyeing versus conventional dyeing of modal fabrics. J Clean Prod 72:127–138. doi:10.1016/j.jclepro.2014.02.002

Tchinda JBS, Pétrissans A, Molina S et al (2014) Study of the feasibility of a natural dye on cellulosic textile supports by red padouk (Pterocarpus soyauxii) and yellow movingui (Distemonanthus benthamianus) extracts. Ind Crops Prod 60:291–297. doi:10.1016/j.indcrop. 2014.06.029

UNEP/SETAC Life cycle initiative (2011) Towards a life cycle sustainable assessment—Making informed choices on products. United Nations Environment Programme (UNEP), Paris, France

Villela A, van der Klift EJC, Mattheussens ESGM et al (2011) Fast chromatographic separation for the quantitation of the main flavone dyes in Reseda luteola (weld). J Chromatogr A 1218 (47):8544–8550

Weidema BP, Thrane M, Christensen P et al (2008) Carbon footprint a catalyst for life cycle assessment? J Ind Ecol 12(1):3–6. doi:10.1111/j.1530-9290.2008.00005.x

Weidema BP (1998) Multi-user test of data quality matrix for product life cycle inventory data. Int J LCA 3(5):259–265

Weidema BP, Wesnæs MS (1996) Data quality management for life cycle inventories—an example of using data quality indicators. J Clean Prod 4(3–4):167–174. doi:10.1016/S0959-6526(96)00043-1

Weidema BP (1993) Market aspects in product life cycle inventory methodology. J Clean Prod 1 (3–4):161–166

Wong (2016) Communication with Patric C.C. Wong, Fong's national engineering Co., Ltd. China-Italy fashion forum, Keqiao, China, 14 Oct 2016

Yi E, Yoo ES (2010) A novel bioactive fabric dyed with unripe Citrusgrandis Osbeck extract part 1: dyeing properties and antimicrobial activity on cotton knit fabrics. Text Res J 80(20):2117–2123. doi:10.1177/0040517510373633

Youssef YA (2016) Added-value products for sustainable textile developments in Egypt. Paper presented at the IFATCC XXIV International Congress, Pardubice, Czeck Republic, 13–16 June 2016

Yuan ZW, Zhu YN, Shi JK et al (2013) Life-cycle assessment of continuous pad-dyeing technology for cotton fabrics. Int J Life Cycle Assess 18:659–672. doi:10.1007/s11367-012-0470-3

Zerazion E, Rosa R, Ferrari E (2016) Phytochemical compounds or their synthetic counterparts? A detailed comparison of the quantitative environmental assessment for the synthesis and extraction of curcumin. Green Chem 18:1807–1818. doi:10.1039/C6GC00090H

Enzyme: A Bio Catalyst for Cleaning up Textile and Apparel Sector

Lalit Jajpura

Abstract The textile and apparel sector provides employment to huge population across the world and plays a major role in global economy. Although, various toxic chemicals start from fiber cultivation/production to apparel production are used in it. Therefore, there is dire need to replace these toxic chemicals with appropriate eco-friendly sustainable alternatives. Enzymes have great potential to detoxify the whole supply chain by replacing various harmful chemicals being used in textile and apparel sector especially in wet processing. Enzymes are biocatalyst and life cannot be thought of without them. Infact, enzymes are precious gift of nature for sustainability. Detox fashion cannot be thought of without considering the enzymes as they not only substitute various toxic chemicals but themselves are biodegradable, sustainable and work at low energy. Amylase, pectinase, lipase, catalase, cellulase, hemicellulase, protease, laccase, sericinase, etc are extensively used enzymes in textile and apparel wet processing. The foregoing chapter discusses detox fashion via applications of these enzymes in sustainable wet processing of cotton, regenerated cellulosic, bast, wool, silk, etc textile materials. Enzyme also has great potential in production and modification of manmade fiber along with waste water treatment, decolorisation, soil remediation and detoxification. Further the impact of biotechnological advancements in enzymatic application in textiles has been discussed.

Keywords Enzyme · Enzyme preparation · Textile wet processing · Desizing · Scouring · Bleaching · Removal of hydrogen peroxide · Biopolishing · Fading of denim · Bast fibres · Wool and silk · Leather processing · Immobilized enzyme

L. Jajpura (✉)
Department of Fashion Technology, Khanpur Kalan, Sonepat, Haryana, India
e-mail: lalitjajpura@yahoo.com

© Springer Nature Singapore Pte Ltd. 2018
S.S. Muthu (ed.), *Detox Fashion*, Textile Science and Clothing Technology,
DOI 10.1007/978-981-10-4876-0_5

1   Introduction

In past centuries numerous developments took place starting from fiber cultivation or production to garment manufacturing. Although these developments fulfilled the huge demands and fashion needs of increased global population but each step is associated with environmental pollution. In fact, the natural fibers e.g. cotton which seems eco-friendly also creates great threat to the environment due to unnecessary use of huge amount of water, synthetic pesticide and fertilizes whereas manmade fiber needs various toxic chemicals as raw material for production and further associated with dispose off problem due to non biodegradability (Jajpura and Singh 2015). Fashion cannot be completed without embellishment using colors, other furnishing and finish effects. Thus, the textile materials and apparels are pretreated, dyed, printed and finished in textile wet processing steps as per the required aesthetic and functional properties. These operations such as desizing, scouring, bleaching, mercerizing, retting, dyeing, printing, finishing etc., need different types of acids, alkalis, oxidizing and reducing agents, surfactants, dyes and pigments, thickeners, auxiliaries, finishing agents, etc. Most of the used chemicals to date are toxic and harm the environment to great extent. There is dire need to detoxify these aforesaid processes by replacing carcinogenic chemicals with appropriate sustainable alternatives. Enzymes play a vital role as detoxifier by replacing various conventionally used toxic chemicals in textile and apparel industries.

The enzyme technology is truly a primitive art and has been used since centuries in food and alcoholic beverages, tanning of leather, etc. The properties and reactions of enzyme catalysis was first recognized by G.S.C. Kirchhoff in 1811 and the actual word "catalysis" was used by Berzelius in 1838. The word "enzyme" was proposed by Kuhne in 1878. The hypothesis for modern enzyme chemistry was proposed profoundly by Michaelis and Menten, while the enzyme urease was isolated by J.P. Sumner (Sumner 1926; Michaelis and Menten 1913).

The first enzyme produced industrially was the fungal amylase takadiastase, employed as a pharmaceutical agent for digestive disorder in the United States as early as 1894. Otto Roehm's patented laundry process for clothing via tryptic enzyme additives was announced in 1915. By 1969, 80% of all laundry detergents contained enzymes, chiefly protease. Along with these, additional enzymes such as lipases, amylases, pectinases, and oxidoreductases were used experimentally in the detergent industry (Crueger and Crueger 2000). There was a massive increase in the use of enzymes in detergents between 1966 and 1969 but it collapsed between 1969 and 1970 due to discovered allergic symptoms in workers handling enzymes at the factory level. However, very soon enzyme encapsulation techniques reduced this risk and once again enzyme applications increased in detergent and textile industries (Smith 1996).

In present scenario due to increase in ecological concern, applications of enzymes are increasing globally in food, agriculture, textile and fashion, leather, chemicals, medicine, effluent treatment, bioremediation, energy, etc sectors. Global market for industrial enzymes was approximately about $4.2 billion in 2016 and expected to develop at a compound annual growth rate (CAGR) of approximately 6% over the period from 2016 to 2022 to reach nearly $6.2 billion (Industrial Enzyme Market 2016; Singh et al. 2016).

At present approximately 200 microbial enzymes are used commercially in which only about 20 are produced at large industrial scale amongst known almost 4000 enzymes. The top three enzyme companies, i.e. Denmark-based Novozymes, US-based DuPont (through the May 2011 acquisition of Denmark-based Danisco) and Switzerland-based Roche rule the market and share approximately 75% share of global production (Li et al. 2012).

2 Enzyme

Enzymes are naturally occurring high molecular weight proteins composed of amino acids. The molecular weight of these ferments is very high and is of the order 10^4–10^5. They are capable of catalyzing the chemical reactions of biological processes and hence are known as "Bio-catalyst". Enzymes are found in plants, microbials as well as animals, where they play an important role in the function of cells and can be considered as living catalysts. They belong to the class of microorganisms such as mould, fungi, yeast, algae, viruses, protozoa, bacteria, etc (Othmer 1980; Boyer 1959).

There are specific biochemical catalysts made to take part in a particular reaction, being refurbished in their original form at the end of a particular reaction. Enzymes can bring about hydrolysis, oxidation, reduction, coagulation, decomposition, although the most common reaction of the enzymes used in textiles is hydrolysis. Significantly higher reaction rates at lower energy consumption can be achieved by enzyme catalysts as compared to the conventional catalysts. The major difference of enzyme catalyst from chemical catalyst is that they are very specific in nature for substrate, temperature sensitive, need relatively low energy for activation and are usually active over a narrow range of pH. Ease of handling, storage and disposal makes enzymes a choice catalyst for any processing industry. With the advances in biotechnology, new strains of enzymes in relatively higher concentrations are now available and they are widely accepted in textile fabric and garment processing units. They are considered to be marvelous molecular machines and a supreme gift of the nature for sustainable wet processing of textiles and apparels.

Textile industries commonly use various harsh chemical agents in their different processes like desizing, cotton softening, denim washing, silk degumming, etc. These chemicals, after their use, cause pollution in the form of effluents; some of them are corrosive, which could damage the equipment and the fabric itself. However, with the introduction of enzymatic processes in textiles, the scenario has changed in recent times. Enzymes being natural products are completely biodegradable and accomplish their work quietly and efficiently at mild condition without leaving any pollutants (Shukla and Jajpura 2004; Shukla et al. 2003; Palmer 1981).

2.1 Enzyme Structure

Enzymes are protein molecules composed from amino acids. Protein may be fibrous or globular type. Fibrous type of proteins such as α-keratin in wool, β-keratin in silk, collagen in skin and elastin in ligaments are strong and insoluble in water. Whereas, enzymes are globular type of protein molecules composed from commonly twenty types of amino acids which form peptide bonds with each other. The formed protein chain is folded in specific three dimensional shapes. The enzyme properties are attributed by sequence of amino acids and its folded three dimensional structural shapes. Each amino acid consist an amino group ($-NH_2$), a carboxylic acid group ($-COOH$) and a side group R. The formation of peptide bond by linking of two amino acids is shown in Fig. 1.

There are commonly 20 different types of amino acids as per attached side group R. The peptide linkages amongst the amino acids form primary structure of the protein molecules. The formed primary polypeptide back bones are arranged in α-Helical, β-sheets, random coil, triple helix and β-turns to form secondary structure of the enzyme. These folded polypeptide chains of secondary structures are also arranged in compact three dimensional shapes which are known as tertiary structure of protein. In this protein structure the non polar or hydrophobic amino acids tend to fold in to the central area of a globular protein to exclude water as much as possible. Whereas, the polar and ionized side chains orients themselves towards the outer protein surface to form hydrogen bonds with water. The size, shape, chemical structure, polarity, etc properties of R groups present on amino acid governs main role in formation of tertiary structure. Whereas available strong disulfide bonds in cysteine linkage stabilizes the tertiary structure of protein. Protein molecule of enzyme may be composed of more than one polypeptide chain or subunit of tertiary structure and form quaternary structure of protein. The formed protein structure of enzyme is stabilized by several individually weak numerous noncovalent interactions such as hydrogen bonds, hydrophobic interactions, electrostatic interactions and van der Waals forces (Rodwell and Kennelly 1999; Jenkins 2003).

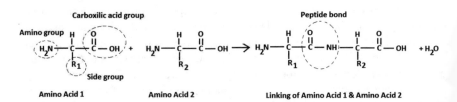

Fig. 1 Formation of peptide bond by reaction of two amino acids

2.2 Cofactor, Coenzyme and Prosthetic Group of Enzyme

All enzymes are proteins. However, without the presence of non protein component called a cofactor, many enzymes lack catalytic activity (Fig. 2). When this is the case, the inactive protein component of enzyme is termed the apoenzyme, and the active enzyme, including cofactor is termed as the holoenzyme.

Cofactors are broadly categorize in following three types:

(1) **Prosthetic group**: an organic cofactor, bound so tightly to the enzyme that it is difficult to remove without damaging the enzyme (Gornall et al. 1949).
(2) **Coenzyme**: an organic cofactor, which is not bound tightly to enzyme and easily removed from the enzyme than a prosthetic group.
(3) **Metal ions**: Metal ions individually may work as cofactor or in certain cases associated with prosthetic groups as ferrous ion in the heme group.

In some cases the cofactors can be removed to form the apoenzyme and be added back later to reconstitute the active holoenzyme. Cofactors fulfill a broad range of reactions in enzymes. One of the main common role of enzyme cofactors is to provide a locus for oxidation/reduction (redox) reaction (Palmer 1981).

2.3 Mechanism of Enzymatic Catalysis

The main ability of enzyme is to bind reactants with themselves. It increases in local reactant concentration and hence in local reaction rate. Beside these enzymes are very much specific for substrate as they can discriminate between different substrate molecules by their particular shape. The specificity for substrates varies for different enzymes. The phenomenon of aforesaid enzyme property can be explained by enzyme specific three dimensional structures as well as active or catalytic site present in it. The enzymes contain true activity centers in the form of three dimensional structures as fissures, holes, pockets and cavities or hollows. This specific portion of enzymes have amino ($-NH_2$, $-NH_3^+$), carboxylic ($-COOH$, $-COO^-$) end groups with R groups containing various amino acid present in the

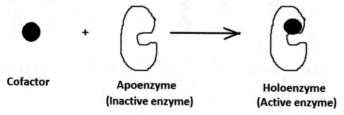

Cofactor   Apoenzyme   Holoenzyme
                (Inactive enzyme)   (Active enzyme)

Fig. 2 Illustration of formation of active holoenzyme from cofactor and inactive apoenzyme

vicinity. The active site is part of the enzyme molecule that combines and reacts with the substrate. The shape and charge distribution of active site help in substrate distinguishing and catalyzing the chemical reaction. The number of active sites per molecule is very small; generally only one. To catalyze a reaction, the enzyme molecule makes a complex by being adsorbed onto the surface of substrate in lock and key fashion (Conn et al. 1987).

This lock and key template model proposed by E. Fisher in 1980, which is still useful for understanding certain properties of enzymes such as specificity and kinetics of a simple substrate saturation curve. The enzyme and substrate interaction on this hypothesis is illustrated in Fig. 3 (Rodwell and Kennelly 1999). This model implies for rigid structure although enzymes are flexible molecule. In 1958 Koshland proposed induced fit model for a substrate which is not exactly fit with the enzyme. In this model, substrate induces a conformational change in the enzyme structure at its active site in such a way that it exactly fit and orient with the substrate. (Cavaco-Paula and Almeida 1996; West and Todd 1957)

Enzymes are biocatalyst and accelerate the rate of chemical reaction without taking direct part in the reaction. The enzymatic reactions clearly requires that the energy of activation be significantly less than that for the corresponding non-enzymatic reaction, as indicated in Fig. 4 (Weil 1996). The symbol ΔE_{WC} shows required activation energy in reaction without catalyst and ΔE_{CC} shows required activation energy in reaction with chemical catalyst whereas ΔE_E shows required activation energy in reaction with enzymes.

Enzymes speed up a particular chemical reaction by lowering the activation energy for the reaction. They achieve this by forming an intermediate enzyme-substrate complex, which alters the energy of the substrate such that it can be more readily converted into the product. For example the decomposition of hydrogen peroxide requires 18 kcal/mole without catalyst and 11.7 kcal/mole in presence of colloidal platinum (chemical catalyst) where as only 2 kcal/mole in presence of the enzyme catalase. One molecule of this enzyme can convert 5×10^6, i.e. five million molecules of hydrogen peroxide to water and oxygen in one minute (Weil 1996).

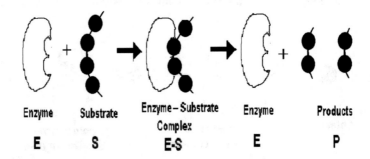

| Enzyme | Substrate | Enzyme–Substrate Complex | Enzyme | Products |
| E | S | E-S | E | P |

Fig. 3 Enzyme and substrate reaction on the basis of lock and key principle

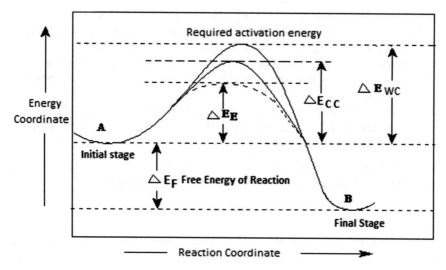

Fig. 4 Enzymes as a catalyst in chemical reaction

2.4 Nomenclature of Enzyme

The Nomenclature Committee of The International Union of Biochemisty and Molecular Biology (NC-IUBMB) classified enzymes into six main classes according to the type of reaction catalysed (NC-IUBMB 1992; Drauz and Waldmann 1994).

The Enzyme Commission proposed that each enzyme should have a code comprising four numbers. The first number defines to which class the enzyme belongs. The main classes of enzymes are further subdivided into subclasses and then sub-subclasses. At the last, enzyme have specific serial number in that particular sub-subclasses. As an example, catalase is an oxidoredutase and is assigned the number EC 1.11.1.6, where first digit 1 denote main class; second digit 11 denote subclass of main class, third digit 1 denote sub-subclass of main class and last fourth digit 6 denote serial number in sub-subclass of main class (Suskling 1984).

There are six classes of enzymes:

1. Oxidoreductases: catalyze oxidation/reduction reactions by transferring hydrogen, oxygen, electron between molecules. e.g. peroxidase, dehydrogenases, oxidases, etc.
2. Transferases: catalyze group of atom such as methyl, phosphate, amino, glycosyl, etc from one donor to another acceptor molecule. e.g. glucokinase, ATP glucose phosphotransferease.
3. Hydrolases: catalyze hydrolytic cleavage of bonds e.g. C–C, C–N, C–O, etc. e.g. amylase, lipase, protease, etc.

4. Lyases: catalyze non hydrolytic cleavage of bonds e.g. C–C, C–N, C–O, etc by elimination reactions leaving double bonds or add groups to a double bonds in reverse reactions. e.g. decarboxylase, pyruvatelyase, aldolase, etc.
5. Isomerases: catalyze geometric and structural changes by isomerisation and transfer reaction within one molecule. e.g. glucose isomerase, maleate isomerase, etc.
6. Ligases: catalyze the covalent joining of two molecules coupled with the hydrolysis of an energy rich bond in ATP or similar triphosphates. e.g. pyruvate carboxyligase (ADP forming).

Most of the enzymes used in the textile industry belong to class 3 i.e. they are hydrolases. Hydrolases catalyze reactions in the following form:

$$A - X + H_2O \rightarrow X - OH + HA$$

The second number in the enzyme code, for hydrolases, describes the type of bond the enzyme hydrolyzes and the third number further defines the reaction catalysed. For example an enzyme with the first two numbers 3.2. is a hydrolase that catalyzes the hydrolysis of glycosidic bonds between carbohydrate residues in polymers such as starch and cellulose.

2.5 Factor Affecting the Enzyme Activity

2.5.1 Substrate Concentration and Enzymatic Activity

The kinetics of enzymatic catalysts can be described by following equation:

$$[S] + [E] \underset{k_1^{-1}}{\overset{k_1}{\rightleftharpoons}} [ES] \rightleftharpoons [ES] \overset{k_2}{\longrightarrow} [E] + [P]$$ Where k_1 and k_2 are rate constants. [S], [E] and [P] are concentrations of substrate, enzyme and product, respectively. The substrate S forms a transitory intermediate complex with the enzyme E as **E-S**. With increase in substrate concentration the enzyme activity increases as curve shown in Fig. 5 (Drauz and Waldmann 1994). When substrate concentration tends towards infinite then the whole enzyme is consumed in formation of enzyme and substrata complex [ES] and the initial speed of reaction tends towards the maximum speed V_{max}.

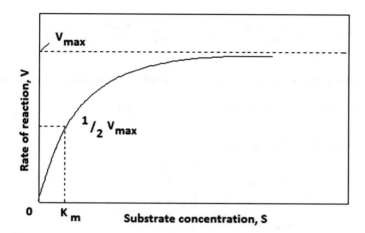

Fig. 5 Effect of substrate concentration on enzyme activity

The **Michaelis–Menten** described the mathematical equation for this reaction in varied substrate concentration as follows:

$$v = \frac{V\max [S]}{K m + [S]}$$

Where, K_m is the Michaelis constant. It is defined as substrate concentration for which the observed rate of reaction is half of V_{max}. The V_{max} shows maximum rate of reaction of enzyme in saturated substrate presence. This rate of reaction can be expressed in another term as catalytic activity of enzyme.

The kinetics measured during a textile process or any enzymatic reaction can provide useful mechanistic information. The Commission of Enzyme has defined an international enzymatic unit called KATAL, defined as the quantity of enzyme, that transforms one molecule of substrate per second under standard conditions of temperature, pH and optimal substrate concentration. Sub-multiples more commonly used are the microkatal (μ kat), the nanokatal (n kat) and the picokatal (p kat) expressed respectively as 10^{-6}, 10^{-9} and 10^{-12} kat (kat). The catalytic enzyme activity is also expressed in International Unit (IU). One IU unit of enzyme catalyze 1μ mole of substrate per minute under specific conditions (NC-IUBMB 1992; Drauz and Waldmann 1994; Suskling 1984).

2.5.2 pH and Buffers

Proteins are composed from various amino acids which contain –NH_2 as well as COOH groups. The presence of acids and alkalis affect the ionization state of these amino and carboxylic acids along with other polar groups in enzymes as shown in Fig. 6. At low pH, the high H^+ concentration converts the amino acid groups

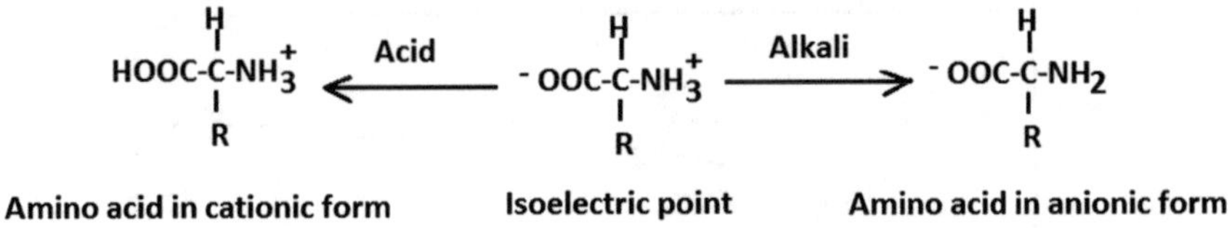

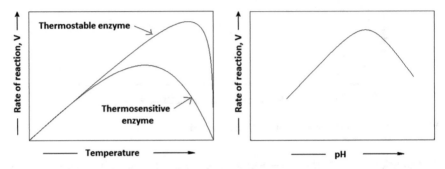

Fig. 6 Effect of pH on end groups of amino acids in enzymes

Fig. 7 Effect of temperature and pH on activity of enzyme

predominately in cationic forms. Whereas, at high pH, the low H^+ (or high OH^-) concentration converts the carboxilic acid groups predominately in anionic forms. The pH at which a protein of enzyme contains equal cationic and anionic groups is known as isoelectric point. Thus even slight change in pH affects the interaction amongst these ionic amino groups which affects the shape, stability and activity of enzyme. Every enzyme has an optimum pH or pH range in which its activity is maximum and beyond this range the activity diminishes as illustrated in Fig. 7. Most of the enzymes optimally work with in pH range 5–9.

2.5.3 Temperature

With increase in temperature, mobility of enzyme and substrate molecules increases which enhance the enzyme activity. But simultaneously, the increased thermal energy also destabilizes the enzyme structure which reduces the enzyme activity. Therefore initially with increase in temperature enzyme activity increases up to optimum point and beyond that with increase in temperature enzyme activity decreases due to thermal denaturation of enzymes. There are certain enzymes which are stable at higher temperature and are known as thermo stable enzymes.

2.5.4 Activator

Many inorganic ions such as Ca^{2+}, Sr^{2+}, Mg^{2+}, Fe^{2+}, Zn^{2+}, etc. are known for increasing the catalytic activity of enzymes and known as activator. These metals stabilize the structure of the enzyme-substrate complex. Being cofactor, certain metals are capable of becoming part of the enzyme and take part in ion exchange reactions (Metzler 2001).

2.5.5 Inhibitors

The enzyme works on substrate by active site which is very specific in shape along with particular charge distribution on its amino acids and other polar groups. Certain chemicals such as alkalis, antiseptics, chelating agents, heavy metals, acid liberating agents, etc may change the confrontation of active site and inhibit the enzyme activity. Even excessive substrate concentration may tend to inhibit the enzyme activity. The enzyme inhibition can be either reversible or irreversible depending on interaction between inhibitor and enzyme active site. The irreversible inhibitors may be also broadly classified as per the mode of action with active site as competitive inhibition, non-competitive inhibition, substrate inhibition, end product inhibition, etc.

2.6 Source of Enzymes

On the basis of source enzymes may be animal, vegetable and microbial based (Shenai 1991).

- **Animal**: The animal based enzymes such as pancreatic enzymes are prepared from slaughterhouse waste such as pancreas, clotted blood, liver, etc. Examples are viveral, novofermosol, degomma, etc.
- **Vegetable**: Malt extracts are made from controlled fermentation of germinated barley. Examples are maltostase, maltoferment, gabahit, diastase, diastafor.
- **Bacterial**: Enzymes are produced by growing cultures of certain microorganisms in sterilized wort. Examples are biolase, amylase, protease, catalase, plurotus ostreatus, etc.

Intracellular or extracellular enzyme: Enzymes can be classified as intracellular and extracellular according to the mode of enzymes or proteins produced or secreted by the bacterial cell. The intracellular enzymes perform their catalytic activity in confined cell membrane whereas extracellular enzymes are secreted outside by the microbial cell into the surrounding medium in which it is living. The intracellular enzymes have open and flexible shape as they are protected in cell's inner environment, thus have less stability towards extreme condition of temperature, pH, salt, etc. Whereas extracellular enzymes have compact and stable shape as they are designed to work in outer environment of cell thus have more stability than intracellular enzymes.

3   Applications of Enzymes in Textile and Apparel Industries

Enzymes are being used extensively in dairy, baking, beverage, animal feed, polymer, paper and pulp, detergent, leather, textile and apparel, cosmetic, pharmaceutical, waste management and effluent treatment, bio diesel, etc. They are one of the best sustainable alternatives to toxic traditional used synthetic chemicals in various processes of textile, apparel and allied industries and play an important role in detox fashion due to the following important properties:

- Enzymes catalyze reactions
- Enzymes have potential to replace numerous toxic chemicals and even exploit for bioremediation
- Enzymes work under optimum mild conditions with less energy requirement
- Enzymes are specific to substrate
- Enzyme reactions are easy to control
- Enzymes are biodegradable and ecofriendly in nature.

Applicable enzymes in detergent, leather, textile and apparel industries are shown in Table 1 (Singh et al. 2016; Shukla et al. 2003; Li et al. 2012). The enzymatic application in textile and apparel sector is discussed one by one in following section.

4   Applications of Enzymes in Wet Processing of Cotton

The preparatory treatments are very important part in wet processing as they make the cellulosic material free from impurities, absorbent and white as per the requirements (Hickman 1995). The escalating energy, labor-cost, stringent pollution control regulations, water scarcity and greater demand for quality textiles have compelled textile chemists to consider preparatory processes more seriously. Enzymes can play pivotal role in detox fashion especially in sustainable preparatory wet processing of cellulosic or other textile materials in processes such as biodesizing, bioscouring, enzymatic bleaching, enzymatic hydrogen peroxide removal, biopolishing or depilling, enzyme assisted dyeing, fading of denim, etc. In following section, need of wet processing step and conventionally used chemicals are briefed initially and thereafter applicable enzymes and their applications have been discussed in detail. It is pertinent to mention that some of the important hydrolytic enzymes are applicable in same as well as other type of category of fibers in various steps with different objectives e.g. cellulases are being used extensively in bio-scouring in conjunction with other enzymes to loosen oily and fatty impurities imbedded in cellulosic polymers. They are also applicable in biopolishing or depilling of cellulosic fibers and fading of denims in biostonewashing. Cellulases more or less play similar type of hydrolysis reaction in cellulosic bast and

Table 1 Applications of enzymes in textile, leather and allied industries

Industries	Enzyme	Applications
Leather	Protease	Dehairing, bating and soaping by hydrolysis of proteins
	Lipase	Degreasing by hydrolysis of lipids and fats
Pulp, paper and raw material preparation for regenerated fibers	Lipase	Deinking and pitch control by hydrolysis of lipid linkages
	Cellulase	Hydrolysis of weak cellulose polymer chains to make the pulp soft and flexible
	Xylanase	Facilitate pulp bleaching
	Laccase	Decolorising or bleaching without use of chlorine and delignification
Textile and garment	Amylase	Removal of starch in desizing without any harm to fabric
	Lipase	Removal of Lipids in desizing and scouring at mild condition without use of harsh alkali
	Pectinase	Removal of pectines in scouring operation Removal of pectines in retting of bast fibers
	Hemicellulase	Removal of hemicellulose in retting of bast fibers
	Oxidoreductase i.e. Laccase, oxidase	Facilitate bleaching and fading of denim
	Catalase	Removal of hydrogen peroxide from residual bleach liquor
	Cellulase	Hydrolysis of cellulose and facilitate scouring operation of cellulosic fibers Defusing of cellulosic fibers e.g. cotton, viscose, lyocell, jute hemp and remie and improve fabric softness Removal of cellulosic impurities in carbonisation of wool without use of harmful acids Removal of cellulose in retting of bast fibers Fading of denim by loosening of indigo dye
	Protease	Hydrolysis of proteins and facilitate scouring operation of protein fibers e.g. wool Degumming of silk Defusing of protein fibers e.g. wool and silk Descaling and shrinkage proofing

(continued)

Table 1 (continued)

Industries	Enzyme	Applications
Detergent	Amylase	Removal of stains of starch and other carbohydrates
	Lipase	Removal of stains of fat
	Protease	Removal of stains of proteins Removal of fuzz and pills of protein fibers and improve color brightness
	Cellulase	Removal of fuzz and pills of cellulosic fibers and improve color brightness of fabrics
	Cutinase	Removal of stains of triglycerides
Effluent treatment and waste management	Amylase, Amyloglucosidase	Bioremediation of starch or vegetable waste
	Lipase	Degradation of hydrocarbon based crude oils or other vegetable lipids
	Nitrile hydratase	Degradation of nitriles containing wastes
	Laccase	Degradation of waste containing olefin, polyurethane and phenolic compounds and specified dyes in effluent
	Manganese peroxidase	Degradation of phenolic compounds and specified dyes in effluent
	Lignin peroxidase	Degradation of phenolic compounds and specified dyes in effluent
	Oxygenase	Degradation of halogenated contaminants and specified dyes in effluent
	Cutinase	Degradation of plastics, Polycaprolactone and specified dyes in effluent

regenerated fibers in different processing steps. Beside cellulosic fibers, cellulases are also being used in wool for removal of cellulosic impurities. Therefore in forgoing chapter textile wet processing operations and enzymatic applications are discussed one by one on the basis of substrate and process sequence for ease of understanding instead of discussing the particular enzyme at one place with their application in different processes and for different textile substrates.

4.1 Enzymes in Desizing

The warp yarns are subjected to various stresses during weaving in a loom due to static or cyclic stretching and friction of fiber and the yarns against the metallic parts such as whip roll, lease rods, wires, etc. The high stress causes breakage of warp yarns. Therefore, it is necessary to prepare good beams for weaving by sizing the warp yarns. The sizing paste consist starch (amylose and amylopectines), poly-acrylates, carboxy methyl cellulose, PVA, lubricant such as softening oils, veg-etable oil, preservative, etc. (Feitkenhauer et al. 2003). The size composition varies according to type of fiber composition in yarn (Khandual et al. 2004). It is necessary to remove the size from the cloth, otherwise the hydrophobicity of the wax and tallow hinder the subsequent dyeing and printing operation (Weil 1996).

Conventionally, size material is removed from the fabric in desizing operation by help of hydrolytic or oxidative chemicals (Batra 1985). Acid used in hydrolytic desizing hydrolyzes the polymeric cellulosic chains resulting in weakening of the textile materials is a main disadvantage of this conventional process. This con-ventional process can be replaced by enzymatic desizing.

4.1.1 Enzymatic Desizing

Use of enzymes in desizing was established decades ago, only in recent years their applications have widened. Enzymatic desizing is performed by application of amylase alone or in combination with lipase (Shukla and Jajpura 2005). Both these enzyme are discussed as below.

- **Amylase**

Amylases (E.C. 3.2.1.1) constitute a class of industrial enzymes, which alone form approximately 25–30% of the enzyme market. The crude industrial amylase consist mixture of α-amylase, β-amylase, glucoamylase and α-glucosidase which hydro-lyze the starch as illustrate in Fig. 8 (Richardson and Gorton 2002). Amylase contains large proportion of tyrosine and tryptophan in the enzyme protein (Shenai 1991). The α-amylase (1-4-α-glucan-glucanohydrolase) degrades starch to oligo-sachcharides and further to maltose and glucose by hydrolysing the α-1,4-glucosidic bond at random. β-amylase (α-1-4-glucan-maltohydrolase) is an exoamylase which hydrolyze α-1,4-glucan linkages of starch chains at the non reducing ends to maltose units. Glucoamylase (α-1-4-glucan glucohydrolase) and α-glucosidase splits starch into glucose units from non reducing terminal ends. α-Glucosidase and glucoamylase are essentially distinguished by releasing

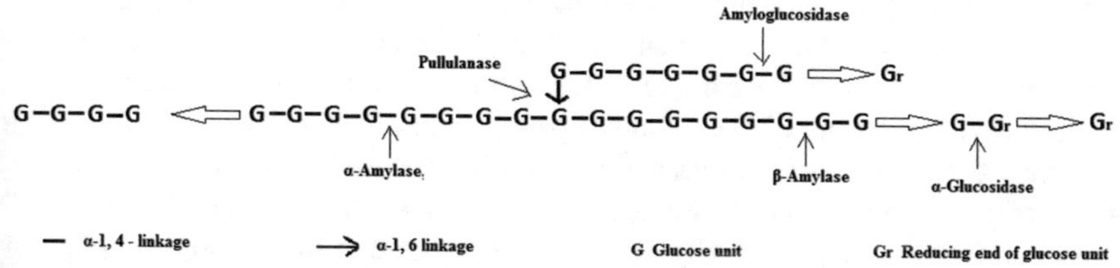

Fig. 8 Illustration of amylase reaction on starch hydrolysis

α-glucose and β-glucose, respectively, from the common substrates having α-glucosidic linkage (Chiba 1997).

Under suitable conditions insoluble starch can be hydrolyzed in soluble starch consisting short chain compounds. The hydrolyzation take place in the following stages (Ajgaonkar et al. 1982).

$$\text{Starch} \rightarrow \text{Dextrin} \rightarrow \text{Maltose} \rightarrow \text{Glucose}$$

The action of enzymes on starch is highly selective and cellulose remains totally unaffected on account of the specific nature of the amylase action (Peters 1967; Mousa 1976). The bacterial and pancreatic amylases consist of a single amylase but the malt enzyme has two constituents α-amylase and β-amylase, which are found to be present in the ratio of 1:5 to 1:6. The results show that α-amylase is more effective than β-amylase in desizing action. The presence of activator such as Ca^{2+} improves the activity as well as stability of amylase. The fats (lipids) used for increasing lubricity in size can be removed by lipase enzymes. A synergism between amylase and lipase is possible that results in even more efficient removal of starch than would occur with amylase alone.

- **Lipase**

Lipases are enzymes, which catalyze hydrolysis of triglycerides into diglycerides, monoglycerides and free fatty acids. Lipases are enzymes with general interest within many industrial applications such as textiles, food, detergent etc (Hasan et al. 2006, 2010; Svendsen 2000).

Lipases hydrolyze the ester bonds in triglycerides, which are a major component of fats. The illustration of catalyzed reaction is as follows:

Lipases in general requires interfacial activation at lipid and water interface. Lipase proceed hydrolysis reaction with the help of Ca^{2+} ions. In desizing operation lipases are used with amylase enzyme to assist the removal of lubricants present in size and enhance absorbency of fabric for better dyeing. Beside desizing, lipase enzymes have great applications too in scouring of cotton, wool and other fiber as well as in detergent, etc. (Varanasi et al. 1997). Alkaline lipases can be effectively used to enhance the removal of oily soils from fabrics and hard surfaces as well as to increase the detergency of commercial products, especially at low temperatures (Xia et al. 1996).

In desizing operation, initially fabric is impregnated with appropriate recipe of amylase and then incubated till size degradation take place and further fabric is washed with alkaline detergent solution at maximum possible temperature to remove solublized starch.

An amylase enzyme can be used for desizing processes at low-temperature (30–60 °C) and optimum pH is 5.5–6.5 (Cavaco-Paulo and Gübitz 2003). Beside desizing, α-amylases are used for the liquefaction of the cereal starch during the production of high fructose corn syrup or fuel ethanol. In household laundry, α amylases present in detergent remove the starch containing stains from clothes and dishes.

4.2 Enzymes in Scouring

The impurities of raw cotton range from 4 to 12% on the weight of cotton. These include waxes, proteins, pectins, ash and miscellaneous substances, such as pigments, hemicelluloses and reducing sugars. Purification of cotton in the fiber, yarn and fabric forms through removal of such impurities is a must prior to its utilization. In the scouring process, the absorbency of the fabric is increased by removal of aforesaid impurities i.e. waxes and fats from cuticle along with pectins, proteins and organic acids from primary cell wall and lumen. The dirt, dust, husk, etc impurities are also removed from the fabric (Trotman 1968; Traore and Buschle-Dilleer 2000).

In conventional method of scouring fabric is boiled in alkali (lime or caustic soda) at atmospheric or at high pressure in Kier machine. Under this process, oil is removed by saponification and wax is removed by emulsification process, respectively. This process consumes huge amount of toxic alkali (approx. NaOH up to 80 kg/ton fiber material), energy and large amount of water for neutralization. Thus, have high BOD, COD load in effluent. These problems can be solved by ecofriendly bioscouring.

4.2.1 Enzymatic Scouring or Bio-scouring

Environmental friendly bio scouring using suitable enzymes is gaining more and more acceptance. Pectinases (Ibrahim et al. 2004), xylanases, cellulases (Li and Hardin 1998), protease (Karapinar and Sariisik 2004), lipases (Sangwatanaroj and Choonukulpong 2003) and pectinases were investigated for their effectiveness in scouring and was found that pectinases are the most effective (Buchert et al. 2000; Tyndall 1992).

- **Pectinase**

Pectins are linear or ramified homo- and hetero-polysachcharides present in plant cell walls as a part of the middle lamella and consist mainly of galactoronic acid (Araujo et al. 2008). In pectin chains, pectic acid are esterified with methyl alcohol and the free hydroxyl group may be also acetylated. Pectinases have been found to be most suitable in bio-scouring as they are capable of removing impurities from raw cotton substrate without adverse changes in the property of the cotton substrate. They are also applied in separating the fibers and eliminating pectins in jute, ramie and flax processing.

Pectinases capable of hydrolyzing pectinic substances at cuticle in raw cotton are enzyme complexes containing mixture of esterases and depolymerases with random or terminal activities (Sakai et al. 1993). The major constituents of pectinases are pectin esterases, polygalacturonases and polygalacturonate lyases. Pectinases (Ahlawat et al. 2009; Hartzell and Hsieh 1998; Buchert et al. 2000) depolymerize the pectins to low molecular water-soluble oligomers and thereby improve the absorbency and whiteness of the textile material without affecting the cellulose back bone and its strength (McNeil et al 1984).

- **Cellulase**

Cellulase enhances effect of pectinase and increase softness of cotton fabric. They often accompany pectinases in small amounts. In scouring, cellulase hydrolyzes cotton removing non-cellulosic impurities effectively but causes damage to fabric that cannot be completely eliminated. Commercial cellulases currently used in the textile industry are usually crude mixtures consisting of a multiple enzyme system, which hydrolyzes cellulose as illustrated in Fig. 9 (Richardson and Gorton 2002). Although a large number of microorganisms are capable of producing cellulase, only a few are producing cellulase in significant quantity, which is capable of completely hydrolyzing crystalline cellulose (Mojsov 2012). Commercial cellulases are mainly produced by fungi such as Trichoderma reesei and Humicola insolens. The crude mixture secreted by Trichoderma reesei fungi consists of following three major types of cellulases (Sarkar and Etters 1999; Etters 1995):

- 1,4-ß-D-glucan 4-glucanohydrolases (endoglucanases) randomly hydrolyze cellulose chain
- 1,4-ß-D-glucan cellobiohydrolases (exocellulase or cellobiohydrolases) split cellobiose from the chain end

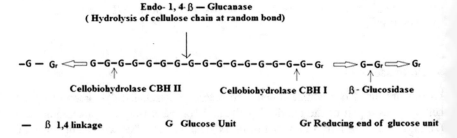

Fig. 9 Illustration of cellulase reaction on cellulose hydrolysis

- ß-D-glucanohydrolases (cellobiases or ß- glucosidases) split cellobiose to glucose.

• Lipase

Waxes and fats impurities are more difficult to remove from cotton fabric. These are hydrophobic in nature. These are complex hydrophobic substances composed of high molecular weight alcohols and fatty acids in free and esterified form. Many lipases are capable of cleaving various natural lipids and oils present inside the cell walls of cotton. The lipase enzyme has been already discussed in length in desizing process.

• Xylanase

Cellulase fibers also having traces of Hemicelluloses which contains xylose, mannose, rhamose, hexose, etc. A portion of hemicelluloses is water-soluble and others bound to micro fibrils. Xylanases in combination with cellulases are capable of cleaving β-1,4 linked xylosyl sequences. Recently the interest in xylanase has markedly increased due to the potential industrial uses, particularly in pulping and bleaching processes (Dhillon et al. 2000). Xylanases have significant applications in production of ethanol, aroma, fruit juices, animal feed, baking, textile, pulp and paper industries.

• Cutinase

As similar to lipase, cutinase is also lipolytic enzyme. Cutinases also hydrolyze the ester bonds of triglycerides and cutin present in waxy layer present in cotton fiber. Although cutinases do not requires interfacial activation and presence of Ca^{2+} as required by lipases enzymes. The cutinases are capable to hydrolyze the cotton waxes at low temperature (Sakai et al. 1993).

Various researchers investigated cellulases, pectinases, proteases and lipases of different origins in their effectiveness to improve absorbency and whiteness in bioscouring operation (Agrawal 2005). The results show that cellulase and pectinase shows maximum weight loss in the fabric with satisfactory improvement in absorbency. Enzymatic bioscouring reduces the effluent load significantly as the Biological Oxygen Demand (BOD) and Chemical Oxygen Demand (COD) of enzymatic scouring process are only 20–45% as compared to alkaline scouring whereas Total Dissolved Solid (TDS) of enzymatic scouring process is 20–50% as compared to traditional alkaline scouring. Although, the bigger size of enzymes as compared to conventionally used alkali limits its penetration in interiors of cellulosic structure. This problem can be minimised by using appropriate surfactants (Mojsov 2011). An LCA study also shows that enzyme application in scouring impacts less on energy, water, and chemicals compared to chemical scouring (Nielsen et al. 2009). The observed advantages of enzymatic bioscouring (Herbots et al. 2008) are as follows:

- Provide soft handle without significant loss in tensile strength.
- Less weight loss in an optimized enzyme process.
- Safer working environment and no corrosive condition for equipments.
- Safe process for blends consisting alkali sensitive fibers.
- Process can be combined with desizing or bio-polishing operation.
- Considerable savings in time, energy, water and use of chemicals.

4.3 Enzymes in Bleaching

Decolorisation of natural pigments present in textile material is carried out by oxidizing or reducing agent in bleaching operation (Hedin et al. 1992). Various chemicals such as sodium chlorite, sodium chlorate, sodium bromite, bleaching powder, hydrogen peroxide, sodium perborate, etc. are used as oxidative bleaching agents. Over the last few years, hypochlorite and other halogen liberating bleaching agents are being replaced by other safer bleaching agents like hydrogen peroxide (Rott and Minke 1999). The chemical oxidative agents are not specific to natural pigments but also oxidize and deteriorate the cellulosic textile material. Recently few enzymes were found suitable for bio bleaching.

4.3.1 Enzymatic Bleaching

Enzymatic bleaching by laccase for the first time was reported in 2003 (Tzanov et al. 2003a, b). It was reported that laccase isolated from strain of *T. hirsuta* improve the whiteness of cotton by oxidizing the flavonoids in it (Pereira et al. 2005). The results of the study show that the laccase can be also employed in improving and stabilizing the whiteness of already bleached fabrics (Couto and Toca-Herrera 2006).

- **Laccase**

Laccase is a cuproprotein belonging oxidoreductase enzyme. It contains four copper atoms which play important role in enzymatic reactions (Eggert et al. 1996). Laccase works on substrate by the laccase mediator system (LMS) as shown in Fig. 10 (Banci et al. 1999). Laccase active site interacts with small molecules of mediator and oxidized them into strong oxidizing intermediate form (co-mediator). Simultaneously laccase active site also interacts with oxygen molecule present in nearby environment as cosubstrate. The oxidized mediator can diffuse away from the active site of enzyme and able to oxidize any substrate which due to size and shape itself directly cannot interact with enzyme's active site. Therefore, laccase (E.C. 1.10.3.2) is quite unspecific to substrates contrary to general enzymes and able to catalyze oxidation of various aromatic compounds (particularly phenols).

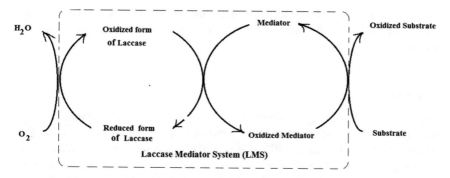

Fig. 10 Illustration of laccase Mediator System

More than 100 possible mediator compounds have been reported but the most commonly used are the azine 2,2'-azino-bis-(3-ethylbenzothiazoline-6sulfonic acid) (ABTS), benzo triazole (BT) and 1-hydroxybenzotriazole (HBT). Considerable resources were exploited in the quest for a cheap, non-toxic mediator that could be used on an industrial scale (Claus 2003). Besides bleaching, laccase has also applications in fading of denim, bast fiber processing, paper and pulp industries as well as in dye degradation and bioremediation (Jajpura 2014).

4.4 Enzymes in Removal of Residual Hydrogen Peroxide After the Bleaching Operation

Hydrogen peroxide is a powerful oxidant used as a bleaching agent or bactericidal in the paper, food, textile and apparel industries. There is continuous increase in application of hydrogen peroxide as preferred bleaching agent as it imparts durable whiteness with less adverse effect on cellulosic fabrics. Its other advantages are simplicity in use, cost effectiveness and comparatively eco-friendly nature than other chemical bleaching agents. However, H_2O_2 is a very toxic substance hence it should be decomposed before disposal (Michael and William 1997). Beside these, no peroxide residue should be left on fabric as it may deteriorate the fabric and change in shade of many sensitive dyes towards oxidation. Traditionally, residual peroxide is removed either by number of rinsing steps or by reduction with inorganic salts followed by rinsing with water. In chemical reduction process, proper dose of reducing agent is very important as lower and higher dose of reducing agent may adversely affect the subsequent dyeing operation (Shukla and Maheshwari 2002).

Alternatively, after the bleaching process, the residual hydrogen peroxide can be removed by addition of catalase enzyme, which is ecofriendly in nature and saves water, energy and time.

- **Catalase**

Catalase is an abundantly available natural enzyme, which decomposes hydrogen peroxide to water and oxygen. The reaction rate is extremely fast: 1 mol of the enzyme is able to decompose 500 million-mol of H_2O_2 in 1 min. In addition to the protein part of the molecule the catalase contains a non-protein part, which is a derivative of haem and includes the metal iron. Catalase protects living organisms against injury by the hydrogen peroxide produced through normal metabolic reactions, or from accidental side reactions (Maehly and Chance 1954; Chance et al 1979).

4.5 Enzymes in Biopolishing or Depilling of Cellulosic Fabrics

Protrude short fibers present on surface of cellulosic fabrics such cotton, viscose, lyocell subdue luster and brilliancy of color and print of the fabric. Further, the short protrude or loose fibers entangle with each other and form pills on the fabric surface during its use and washing treatment. These pills are not required as it give used look and dirty appearance to fabric. The fuzzy protruding cellulose fibers and microfibrils on the fabric surface can be removed efficiently with cellulase enzyme to get smoother and brighter fabric surface with reduced pilling propensity. In enzymatic treatment industrial cellulase enzyme decomposes the cellulose polymeric chains by hydrolytic reaction. Cellulosic polymers present on fabric surface as well as on protrude fibers are maximally exposed to enzyme treatment hence with appropriate hydrolysis, insoluble cellulosic polymer soublized and leach out in solution of reaction bath (Morgado et al. 2000). The optimized enzymatic recipe with prolong time and at higher temperature can facilitate enzyme penetration to the interior of the fibers and result in more severe hydrolysis (Schimper et al. 2009). The increase in hydrolysis enhances the pilling panel value (1—severe pilling, 5— no pilling). However, hydrolysis of cellulosic material in terms of weight loss should not be more than 3–5% as weight loss beyond that affects the fabric strength adversely (Rau et al. 2008).

The biopolishing treatment can be performed on yarns, fabrics or garments before, after or during dyeing treatment or even during home laundering (Snyder 1997). It was observed that the introduction of agitation in enzymatic bath increases the process yield. Besides, applications of certain surfactants and auxiliaries also enhance the efficiency of biopolishing (Lenting and Warmoeskerken 2001). The biopolishing treatment improves pilling resistance, cleaner surface, drapability, suppleness, softness, brightness of the colored and printed surface along with reduction in dead and immature fibers.

Biopolishing of cotton, viscose, linen and lyocell was studied with cellulase enzyme and it was found that linen was most susceptible to enzymatic hydrolysis followed by viscose rayon, cotton, and lyocell. The regenerated man made fiber lyocell is associated with fibrillation which can also be removed by treatment with cellulase (Kumar et al. 1997).

4.6   Enzymes in Fading of Denim

Denim is one of the world's oldest fabrics and commonly associated with jeans. A denim jean was invented nearly 150 years ago by Levi Strauss. At present denim is fashion trendsetter and its focus on durability has changed to fashion. Denim is known for its faded vintage look. Denim is faded by various means such as by various chemicals, ozone treatment, plasma treatment, laser beams and abrasion by sand blasting, ceramics and sand paper, etc.

Traditionally denim fabric is desized by acids or other chemicals for requisite softness and further fading is performed by acids and oxidative bleaching agents by which decolorization of dyes take place as per the fashion trend. These chemical agents are applied on whole fabric by exhaustion method or by sponge at localized place. In acid wash treatment, various types of acids alone or in conjunction with other chemicals are used to hydrolyze the cellulosic polymer at the fabric surface so that loosely held dye particles may leach out and consequently faded effect may be obtained. These aforesaid chemicals used for fading are not specific only to fading but also deteriorate the textile material resulting in reduced fabric strength.

The denim garments are also faded in stone wash operations in which garments are laundered with pumice stone alone or in combination with other chemicals. The rotation of garments with pumice stone causes abrasion on garment surface resulting in fading. The stone wash process has following demerits:

- Wear and tear in the garments, their seams and accessories
- Handling problem of pumice stone and its formed residue
- Deposition of pumice stone residue on to the fabric surface
- Requirement of huge amount of water and energy along with labor intensive process.

The all aforesaid fading treatment with pumice stone and hypochlorite were followed by neutralization and a rinsing step, all causing substantial environmental pollution (Pedersen and Schneider 1998).

- **Enzymatic Fading of Denim with Cellulase**

The cellulase enzyme as discussed earlier in scouring and biopolishing of cellulosic section also has application in fading of denims. Cellulase leaches out cellulose polymers from the fabric surface by hydrolyzing the cellulose polymer chains. The indigo dye particles present on the exterior surface of ring dyed warp yarn are loosened in the cellulase treatment and washed away resulting in fading effect. The process is known as "bio-stonewashing". A small amount of cellulase in bio-stonewashing process can replace several kilograms of pumice stone. Cellulase enzyme alone or with small amount of pumice stone can create various types of finishing effect with providing desirable softening effect to the garment.

Back staining: The leached out indigo dye in the enzymatic bath work as dyeing bath and redeposit on fabric resulting in staining of undyed white weft yarns and white fabric used for pockets and other trims. This phenomenon is known as **back staining** and depends on type of cellulase, its affinity for the fabric and the indigo dye, auxiliary and pH of enzymatic reaction bath (Cavaco-Paulo et al. 1998). On the basis of optimum activity exhibited cellulase are classified as acid, alkali and neutral cellulase and each type of cellulase provide unique type of finish to denim garment. The acid cellulase works at optimum pH 5 and shows maximum fading amongst **acid, alkali and neutral cellulase**. Although, it is also associated with maximum back staining. The neutral cellulase works at optimum pH 7 and shows less problem of back staining.

- **Enzymatic Fading of Denim or Bio Bleaching with Laccase**

The laccase oxidation techniques have potential within a great variety of industrial fields such as pulp and paper, textile and leather, effluent treatment, bio remediation and food industries (Tzanov et al. 2003a; Campos et al. 2001). Laccase is used in commercial textile applications to improve the whiteness in conventional bleaching of cotton and wood pulp in regenerated fiber production and processing of bast fibers. Lacasse has good potential to decolorize the indigo dye. Therefore it has applicability in fading of indigo dyed denims. There is no affect on denim strength as contrary to cellulase, laccase only degrade indigo dye without affecting the cellulosic polymeric chains. The traditionally used hypochlorite bleaching agents are although cheap but affect the environment and denim fabric to great extent. Besides these, hypochlorite also affect the elasticity of elastomeric yarn within the stretchable denims used now a days. Therefore, laccase can be alternatively used alone on denims or further brightening up the denim after cellulase treatment.

Novozyme (Novo Nordisk, Denmark) launched industrial laccase enzyme DeniLiteTM in 1996. Further, DeniLiteII TM based on a new type of laccase with higher activity than that of DeniliteI TM was launched. Similarly Zytex Pvt. Ltd.,

Mumbai, India developed formulation for denim fading in 2001 as Zylite as the trade name of the product (Couto and Toca-Herrera 2006).

5   Applications of Enzymes in Wet Processing of Bast Fibres: Retting, Softening and Defuzzing

Bast fibers such as jute, flax, hemp, ramie, kenaff (Hibiscus sativa), and coir from coconut husks are multicellular vegetable fibers (Morton and Hearle 1993). They are generally composed of cellulose, hemicelluloses and aromatic lignin polymer.

The major composition of few important bast fibers are as follows (Mather and Wardman 2011; Mohanty et al. 2005):

Jute fibers: cellulose, 61–71%; hemicelluloses, 14–20%; lignin, 12–13% and pectin, 0.2%.
Ramie fibers: cellulose, 91–93%; hemicelluloses, 2.5%; lignin, 0.65% and pectin, 0.63%.
Flax fibers: cellulose, 60–70%; hemicelluloses, ≈17%; lignin, 2–3%; pectin, 10% and wax 2%.
Hemp fibers: cellulose, ≈77%; hemicelluloses, 0–1%; lignin, 1.7%; pectin 1.4% and wax 1.4%.

These fibers are integral part of the stem structure and before spinning they have to be separated from other tissue of the woody plants in retting or degumming operation. Traditionally, retting of bast fibers such as jute, flax, or hemp can be performed by water, dew and chemical. In water retting the plants straw is submerged in water pits or running fresh water in which due to fermentation certain bacteria and fungi decompose the pectin of bark and release fiber in anaerobic process (Trotman 1984). Whereas in dew retting, wet straw is exposed in moisten condition which caused to incubation of fungi and aerobic bacteria. The anaerobic water retting in ponds is associated with numerous problems such as excessive pollution in ponds or water ways, drying of retted stems and putrid odor. Therefore aerobic dew retting replaced the water retting in mid 20th century. But dew retting process is also time consuming, weather dependant and associated with uncontrolled bacterial or fungal growth along with inconsistent fiber quality. Chemical retting consists softening of plant straw by the application of acetic acid or alkali. (Chesson 1980; Brühlmann et al. 1994). In case of Remie, the fibers are removed from stalks mechanically in decortication process. The decorticated ramie fibers contain 20–30% pectin and hemicellulose as gummy substances. These gummy materials are removed traditionally by degumming process by hot alkali solution. These all aforesaid retting and chemical degumming process requires high energy inputs and toxic chemicals (Das Gupta et al. 1976). These traditional retting and degumming of bast fibers can be replaced by faster more reproducible ecofriendly enzymatic processes with pectinases, hemicellulases, etc enzymes (Brühlmann et al. 2000).

- **Hemicellulases**

Cell walls of land plants (bast fiber plants such as flax, jute and ramie) consist of large amounts of cellulose microfibrils embedded in a continuous phase, of which hemicellulose and pectins are predominant (Akin et al 1999, 2000). Pectinases and hemicellulases have been used under controlled conditions in the retting of flax fibers. This treatment has proven to be quicker and more environment friendly than the traditional retting.

The fungi Aspergillus Sp. and Aspergillus Niger are the main commercial sources of hemicellulases. However, significant hemicellulase activity is present in many cellulase and pectinase preparations. Mixture of commercial enzymes such as polygalacturonases, pectin lyase, hemicellulase and cellulase were screened and developed for retting of bast fibers. Novozyme 249 and Flaxzyme products of Novo reported to produce good quality and yield of retting. Although continual present activity of cellulase enzyme causes significant strength loss of the fiber which necessitates denaturing the cellulase enzyme with alkaline hypochlorite (Van Sumere and Sharma 1991). Development of cellulase free pectinolytic enzyme will be definitely beneficial for getting better yield and quality of retted bast fiber with minimal strength loss.

6 Applications of Enzymes in Wet Processing of Wool

Wool like all other animal fibers are composed of keratin, which is high sulfur content protein. Raw wool contains around 30–70% impurities such as suint (residue of perspiration), dirt (soil particles and cellulosic impurities such as straw, seed, burrs of bast fiber, etc), fat (higher fatty acids derivatives in waxy form) and mineral matters (Trotman 1984). These impurities are removed in traditional processing steps by various chemicals and auxiliaries which are not eco friendly in nature. The various traditional chemical processes and their suitable alternative enzymatic processes are discussed herewith. The commonly used enzymes on wool are cellulase, pectinase, lipase, papain, trypsin, lipoprotein, keratinase, etc.

6.1 Enzymes in Carbonization of Wool

The cellulosic impurities are traditionally removed by carbonization process by use of concentrated sulphuric or other inorganic acids at high temperature. The chemicals used in aforesaid preparatory wet processing are not environment friendly. Thus instead of acid, these cellulosic impurities can be efficiently removed by the use of cellulase, pectinases, ligninases enzymes in appropriate combination without any chemical and physical loss to wool (Heine et al. 2000).

6.2 Enzymes in Scouring of Wool

The impurities such as suits, fats, particulate soil particles are removed in traditional scouring by use of detergent in presence of weak or no alkali or by using solvents. The alkali used in traditional scouring is toxic and harms the wool keratin where as solvents may be toxic and costly. Thus enzymatic scouring with the help of protease reduces requirement of such toxic chemicals as well as water and energy. It has been observed that protease scoured wool fabric gives improved whiteness index and dye ability in subsequent bleaching and dyeing operations.

- **Proteases**

Protease is used in wool and silk processing industries and after amylase is the second largest industrial enzyme (Yoon et al 2000). Commercially protease is produced both from bacteria and fungi. Proteolytic enzymes or proteases catalyze the hydrolysis of certain peptide bonds in protein molecules. Different proteases have different specificity in regard to which peptide bonds are broken. As per the optimum pH range of working, proteases are categorized as alkaline, neutral and acid proteases. Alkaline protease is mainly used in detergent industry as it is stable at high temperature and wide alkaline pH range from 9 to 11. On the basis of chemical reactions, proteases may be classified as serine protease, cysteine protease, aspartic protease and metallo protease. Mainly protease is used in bio processing such as scouring, shrink proofing (descaling), defuzzing of wool and degumming of silk (Rinsey and Karpagam 2012). Bioscouring of wool by protease alone or in combination with lipase can replace traditionally used chemicals with efficient results.

6.3 Enzymes in Finishing of Wool

The cuticle scales present on wool fiber surface causes various problems such as directional frictional effect, uneven shrinkage, felting as well as itching sensation to wearers. Therefore, wool needs surface modification. In traditional method scales are smoothen by hypochlorite treatment in combination with Hercosett or Nopcobond resins or coating with other polymeric substance (Van Rensburg and Barkhuysen1983). Potassium permanganate is also being used effectively in elimination of felting and to impart shrink resistance (Bahi et al. 2007). In recent developments, physical modification of wool were reported efficiently by plasma and UV/Ozone treatments (Shao et al. 1997; Ceria et al. 2010). These treatments induce functionalization and etching by oxidation reaction on exposed surface of wool fiber. Although chlorination and other aforesaid treatments need either toxic chemicals or required high energy resources.

The enzymatic protease treatment can hydrolyze scaly surface of outer cuticle layer in sustainable way. The bio scoured and sodium bisulphate pre reduced wool shows good result of anti shrinkage treatment when treated with protease enzymes. The protease treatment removes the scales of wool fiber as well as whole protruding fibers resulting in smoother, softer wool with less tendency to shrink, felt or pilling formation in subsequent use. The laccase in presence of appropriate mediator can also facilitate the anti-shrinkage treatment of wool (Lantto et al. 2004).

6.4 Enzyme-Mediated Cross Linking of Wool

The treatment with oxidative chemicals and protease causes 10 to 18% strength loss. The transglutaminase enzyme has potential to resume the strength of wool fabric by 3–5% by acyltransfer reaction between glutamine and lysine with the formation of carboxylamide groups of peptide-bound glutamine in wool keratin (Cardamone 2007). Transglutaminase treatment also imparts a significant reduction in fabric shrinkage. Transglutaminases applications on wool indicates that a number of novel and radically different finishes for wool textiles can be developed further (Cortez et al. 2004).

7 Applications of Enzymes in Wet Processing of Silk

The silk filament is secreted by the silkworm Bombyx mori. In raw silk two proteineous fibroin filaments are surrounded by a gummy cementing layer known as sericin. The fibroin is crystalline in nature and around 75% in composition whereas sericin is amorphous proteineous biopolymer and around 25% in composition. The raw silk cocoons are processed to the finished clothing articles by series of steps which include reeling, weaving, degumming, dyeing, printing and finishing operations (Zahn 1993).

7.1 Enzymes in Degumming of Silk

The natural silk filament consists gummy material (mainly sericin) on its surface, which subdues its luster. Conventionally sericin is removed from silk in degumming procedure by the help of alkali, acid or soap to give luster to the silk (Rangi and Jajpura 2015). The degumming of sericin took place by hydrolyzation, dispersion and solubilization of different sericin polypeptides. Chemical degumming causes fiber degradation resulting in loss in aesthetic and physical properties, such as dull appearance, surface fibrillation, poor handle, drop of tensile strength, as well as uneven dyestuff absorption during subsequent dyeing and printing (Freddi et al.

2003). The conventional degumming process consumes huge amount of water and energy as well as chemical used and by product formed due to hydrolysis of sericin has great impact on effluents. The enzymatic degumming process solves some of these problems and lowers the environment impact of effluent.

7.1.1 Enzymatic Degumming or Bio Degumming

Proteolytic enzymes such as acidic, neutral, and alkaline proteases are used in hydrolysis of sericin present on raw silk in enzymatic degumming process. It has been reported that alkaline proteases perform better than acidic and neutral ones in terms of sericin removal, improvement in surface smoothness, handle and luster with less effect on tensile strength. The combination of a lipase with protease also results in improvement in removal of fatty acids. It improves wettability and dyeablity of degummed silk fabric (Gulrajani et al. 2000a, b; Freddi et al 2003).

8 Applications of Enzymes in Dyeing and Dye Synthesis

Enzyme can also play role in sustainable dyeing operation. Laccases is able to generate color "in situ" from originally non-colored low-molecular substances. It has been reported that laccase is able to color the hydroquinone padded wool fabric (Shin et al. 2001). Dyeing of wool was also reported by the use of dye precursor (2,5-diaminobenzenesulfonic acid), dye modifiers (catechol and resorcinol) and laccase, without any dyeing auxiliaries (Tzanov et al. 2003b). It has been reported that laccases is able to synthesize red azo dyes by the oxidative coupling of 3-methyl-2-benzothiazolinone hydrazone (MBTH) and phenols (Setti et al. (1999).

The sulfur dyes are insoluble in nature and needs high amount of sodium sulphide for reduction and solubilization. The sodium sulphide and other synthetic reducing agents are highly toxic in nature, produces hydrogen sulphide which increases sulphur content in waste water. It has been reported that oxidoreductase enzyme such as catalase can be employed as alternative to chemical reducing agent in reduction of sulfur dyes (Chakraborty and Jaruhar 2014).

9 Applications of Enzymes in Surface Modification and Decomposition of Synthetic Fibers

The synthetic fibers such as poly ethylene terephtalate (PET), polyamide, acrylonitrile, etc are sharing approximately 50% share of global fiber market. These fibers exhibit good strength, chemical resistance along with good abrasion, shrinkage and thermal stability. The share of these fibers is increasing continuously due to increase in fibers demand and limited resources for cultivation of natural fibers. Although these fibers have very low moisture regain due to lack in hydrophilicity. Further, most of the synthetic fibers are non biodegradable and have disposal problems and great threat for environment and sustainability. To cope up the pile of garbage of synthetic fiber wastes, there is dire need to find out alternative biodegradable synthetic fibers or develop some methods by which these fibers can be deteriorated. There are several techniques such as plasma treatment, chemical hydrolysis, chemical grafting, etc by which hydrophilicity of synthetic polymers can be improved although these techniques are also associated with huge energy, water and toxic chemicals requirement. In certain cases chemical used also change the fibers properties adversely such as yellowing, change in color, loss in strength, etc. Enzymatic hydrolysis of these synthetic polymers is best alternative to already used surface modification techniques. Thus enzymes can play important role in detox fashion by ecofreindly surface modification and waste disposal of synthetic fibers.

Esterases and lipases have capacity to hydrolyze the ester bond of polyester. Cutinases which are acyl esterases can be used in hydrolysis of vinyl acetate present as copolymer in acrylic fibers. (Silva et al. 2005; Yoon et al. 2002). The strain of cutinase was developed which are capable to hydrolyze PET and PA fibers (Silva et al. 2007). The nitrilases are capable to modify the nitrile groups of acrylic fiber in to carboxylic acid groups which improves dye uptake and hydrophilicity of fiber. The nitrile hydratases and amidases also react with acrylic polymers and improves hydrophilicity. These surface modification experiments are still at the lab scale and yet to be industrialized so textile and apparel industries can move forward towards sustainable processing.

10 Applications of Enzymes in Waste Water Treatment, Decolorisation, Soil Remediation and Detoxification

The textile effluents are extremely variable in composition as they contain different types of dyes, various types of chemical auxiliaries, finishes and reaction by products (Jajpura 2015). The use of synthetic organic dyes increased tremendously and it is estimated that more than 10,000 different types of dyes and pigments are produced worldwide. The global consumption of dyes and pigments was around 30 million tones with expected growth at 3% per annum (Walker and Weatherly 1997). These dyes along with other chemicals and auxiliaries been used on textile substrate and their effluent is drained without appropriate treatment to the nearby environment. Release of colored particles in effluents causes serious pollution threat and adversely affects the human health particularly in developing countries (Bizuneh 2012).

Effluent of textile industry is characterized by high chemical demand (COD), low biodegradability, high salt and heavy metal contents, aromatic amines, carcinogenic azo dyes, etc (Alinsafi et al. 2005). Even small traces of color substance present in water prevent the exposure of sunlight to aquatic plant species affecting the photosynthesis reaction. It causes reduction in available dissolved oxygen for aquatic life (Wu et al. 2004). The lack of sunlight, depleted dissolved oxygen concentration with excessive presence of salts, heavy metals and other toxic chemicals are lethal for aquatic life as well as human health.

Removal of colored compounds from textile industry effluents by physico-chemical methods such as adsorption, precipitation or chemical degradation and biological methods is currently available but being very expensive limits their application (Jajpura et al. 2004; Ramachandran et al. 2013).

The capability of laccases to act on chromophore compounds such as azo, triarylmethane, anthraquinonic and indigoid dyes leads to the suggestion that they can be applied in industrial decolorization processes (Damsus et al. 1991; Liu et al. 2004). In recent years, biodegradative abilities of extracellular nonspecific free radical-based enzyme secreted by some white rot fungi was found promising for decolorising the synthetic dyes. The laccases and peroxidases such as lignin peroxidase, manganese peroxidase, chloroperoxidase and horseradish peroxidase by the help of mediators can catalyze the oxidation of a wide range of toxic organic compounds such as polycyclic aromatic hydrocarbons (PAHs), phenols, organophosphorus pesticides, pesticides, toxic amines and azo dyes (Torres et al. 2003).

Although deterioration of hydrophobic substrates need presence of organic solvent, which adversely affect the enzyme catalytic activity. Beside these, it is found that most of the mediators are also toxic and high cost of enzyme and mediator are the significant limitations of this eco-friendly enzymatic effluent treatment which need to be addressed seriously (Torres et al. 2003).

11 Applications of Enzymes in Leather Processing

The meat processing industries generate hides of dead animals which are further processed into leather which are extensively used in apparel and fashion accessories. Although leather industries pollute the environment at large extent in their various processing steps. It is reported that leather industries discharges more than 15,000 million wastewater liters/day along with generation of approximately 6 million tons Solid waste/year and 4.5 million tons sludge/year (De Souza and Gutterres 2012; Rajamani et al. 2009).

The extensively used chemicals in leather processing are lime and sulphide in hair removal and chromium salts in tanning. The formation of hydrogen sulfide (H_2S) during effluent treatment poses a serious environmental problem and lethal for aquatic life. The lipases, keratinases and proteases are being used in enzymatic applications in various sustainable leather processing steps such as soaking, bating, un-hairing and degrasing.

11.1   Enzymes in Bating

In traditional bating operation protein residues are removed from hides by manure of dog, pigeon or hen to improve the pliability of leather. The industrial protease treatment replaces very unpleasant, unreliable and slow methods of traditional bating into cleaner and efficient process.

11.2   Enzymes in Soaking

In soaking process dirt, blood, flesh, grease, dung etc are removed from hides and skins by use of chemicals and surface active agents. Lipases and proteases can efficiently remove these impurities from leather in enzymatic soaking treatment.

11.3   Enzymes in Un-hairing

In un-hairing operation hairs from hides are removed by use of lime and sodium sulphide alkaline reducing agents. Proteases assisted un-hairing process results in a cleaner and softer leather surface. Finding of some research results shows that enzymes and sodium metasilicate in coenzymatic hair removal process can reduce chemical oxygen demand (COD) and total solid loads by 55 and 25%, respectively (Bhavan et al. 2008).

11.4   Enzymes in Degreasing

In traditional degreasing treatment fat is removed from leather by the use of solvent so further efficient tanning and dyeing can be performed. Lipase alone or in combination with surfactant replaces the use of solvents and removes the fats present on hides uniformly and efficiently.

12   Enzyme Preparation

The chemicals and auxiliaries used such as acids, alkalis, surfactants, chelators, dispersing agents, reducing and oxidizing agents, etc in textile wet processing affect the enzyme activity and its stability to large extent (Shukla et al. 2003). The advance enzyme technology has potential to overcome these aforesaid problems up to certain extents in the industrial enzyme production steps. Enzymes cannot be chemically synthesized but extracted from appropriate living species. The industrial enzymes can be preferably produced from microorganisms as there are certain limitations in mass production from animals and plants. The enzymes are produced from microorganisms in the following main steps:

12.1 Selection or Isolation of Microorganism

The advancement in molecular biotechnology made it possible to select and screen the bacteria which produce the enzyme in good yield with requisite properties. Microorganism selection is tedious process and affects the economy of enzyme production as well as requisite enzyme properties such as broader working range of pH and temperature, higher stability towards alkalis, acids and other chemicals, higher enzyme activity, long storage life and low production cost. Beside that, particular substrate and coenzyme specificity, altered pH and temperature optimum, improve folding properties for better stability of enzyme are some of the other requirements. The appropriate selection of microorganism not only provide enzyme with higher yield and stability but also provide good enzyme recovery, independence of inducer. The problem of odor, color and presence of harmful enzymes and other by-product formation can be also minimized by microorganism selection.

Further, microorganism can be manipulated for producing the enzymes with required properties by the help of genetic and protein engineering. The Recombinant Deoxyribonucleic Acid (RDNA) technology made it possible to transfer specific units of genetic information from one organism to another. The genetic modifications offer higher enzyme production efficiency, thus the lesser use of raw material and energy with formation of lesser by product and waste.

12.2 Selection of Growth Condition for Multiplication of Microorganism and Cultivation of Enzyme in Fermentation Process

After selection of microorganism and its genetic modification, industrial enzymes are produced in bulk by multiplying the microorganisms in fermentation process. Fermentation is carried out in solid or in submerged state in the presence of low cost nutrients such as starches and its derivative or other suitable polysaccharides. The solid state fermentation process is carried out generally on rectangular/circular shallow trays at regulated process parameters such as temperature, humidity, agitation, etc. Developments in submerged fermentation technology made it feasible to produce low cost enzymes in batch or continuous bioreactors in bulk amounts.

12.3   Purification in Downstream Processing

Further in downstream processing, enzymes are isolated from broth (nutrients, bacteria and other additive of fermentation media) by the help of extraction and filtration operations. The enzymes are purified by the help of various techniques such as precipitation, centrifugation, spray drying, hammer or ball mill processing, ultra filtration, electrophoresis or other chromatography, etc. Enzyme may be in solid or liquid; purified or crude form as per the requirement.

12.4   Stabilization of Crude or Purified Enzyme

After the downstream processing, enzymes are formulated to get maximum storage and application stability. Enzymes must be stored at temperature as low as possible in air tight container to minimize the enzyme inactivation. Although storage temperature of enzyme in liquid form must not be below the freezing point of water as it may disrupt the enzyme structure. Similarly optimum pH must be maintained by appropriate buffer for storage and operational stability. Beside these other additives such as salts (i.e. sodium chloride, calcium chloride, ammonium sulphate, etc), sugar and surbitols, glycols, glycerols, PVA, etc being added to enhance maximum shelf life of enzymes. There are chances that contamination of protease producing microorganism may affect the enzyme stability. Thus, microbicides are also added in formulation. In certain cases enzymes are chemically modified by bifunctional cross linking agents or grafting to polysaccharides and synthetic polymers.

13   Reuse Applications by Immobilized Enzymes

The enzymes are precious gift of nature for bio-reactions which play efficient role in sustainable industrial applications. In favorable reaction conditions, even after completion of enzymatic process, soluble or free enzyme still retain their catalytic activity. Although, these free enzymes are difficult to recover from the exhausted reaction bath. Even, the exhausted bath cannot be reused further for fresh substrate as presence of already formed product and by product hinders the fresh reaction of enzymes with substrate molecules. Hence, generally free enzymes are drained after completion of enzymatic process in wasteful practice. The immobilization techniques made it possible to use the enzyme repeatedly.

Enzymes are described as immobilized when they are bound or adsorbed to carrier materials as shown in Fig. 11 (Messing 1975). The handling of insoluble carrier materials containing the enzymes are easier for reuse, as enzymes can be separated from the reaction participants by simple procedure. Beside these, enzymes can be entrapped in soluble form such as microcapsules or membrane, that

Fig. 11 Schematic representation of immobilized enzymes

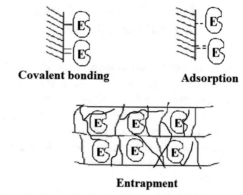

Covalent bonding   **Adsorption**

Entrapment

are impermeable to enzyme, ensuring a continuous exchange of substrate or product. The easy removal of enzyme from the treated product may be a distinct advantage in industry applications as formed product also free from traces of enzymes. Enzyme can be immobilized by following techniques.

13.1 Immobilization by Adsorption

Adsorption is most economical and simple procedure for immobilizing enzymes. The adsorption procedure may not always be the final process of choice as enzyme may leach out from the adsorbing material.

The important factors which affect immobilized enzyme efficacy and stability in adsorption technique are as follows:

1. Ion-ion interaction or hydrogen bonding
2. Ionic strength, temperature and pH of adsorption reaction
3. Effect of adsorption upon enzyme protein or active site
4. Enzyme load on adsorbing material.

13.2 Immobilization by Microencapsulation

The polymerization or coagulation/ precipitation reactions of gel materials in presence of enzyme give entrapped or microencapsulated enzymes. The substrates and products are permeable to membrane formed but the enzyme due to large size cannot be leaked out. Polyacrylamides, calcium alginate, collagen and gelatin, etc are the most commonly used entrapment media. Out of these aforesaid gels, polyacrylamide is the most widely used matrix for entrapping enzymes due its non ionic characteristic which affects the enzyme properties to lesser extent in gel matrix (Bernath and Vieth 1974).

13.3   Immobilization by Covalent Bonding of Enzyme to Insoluble Carrier

Enzymes can be bound to insoluble carriers by various methods. The enzyme must be linked at some distance from the active site of enzyme. Such insolubilized enzymes retain their activities, but their immobility may reduce the reaction rate. The anionic or cationic nature of the carrier may also alter pH optimum for the reaction. The binding of enzyme to carrier may result in steric hindrance and impose restrictions on specificity of bound enzyme. This is most apparent where the substrate is a large molecule, such as protein, rather than a small molecule like a peptide.

Enzymes are covalently bound by help of binding agents such as glutaradehyde, cyanogen bromide, carbobiimide, etc to any suitable inorganic carriers (ceramics, glass, zirconium, magnetic particles, silicates derived of china clay, etc), natural polymers (Sepharose, cellulose, CMC, etc.) or synthetic polymers (nylon, acrylonitrial, etc.). The morphology of carrier such as porous or non-porous is extremely important with respect to surface area and pore parameters, both of which affect the loading of enzyme. Beside this chemical durability, thermal stability, mechanical strength, dimensional stability, microbial resistance, etc properties of carrier plays important role in immobilization (Taylor 1991).

13.4   Properties of Immobilized Enzymes

- Provide feasibility to reuse the enzyme repeated number of times and improves the economy of process
- Ideal for continuous operations in reactor
- No impurity of enzyme in product
- Improve enzyme stability towards temperature, pH and other chemicals
- Provide prolonged stability when compared to the soluble or native form.

13.5   Applications of Immobilized Enzymes

The Immobilized enzymes or whole microbial cells is being extensively used in food and pharmaceutical industries such as immobilized penicillin acylase is used to prepare 6-amino penicillanic (intermediate for penicillins drug), immobilized glucose isomerase is used for industrial production of high fructose syrups from glucose, aminoacylase for production of amino acids, etc.

Textile substrate and immobilized enzyme both are in solid phase thus difficult to interact due to hindrance. This limitation restricts efficient applications of immobilized enzymes in textile wet processing operations such as biodesizing, bioscouring, biopolishing, etc. Although the immobilized enzymes and microbial cells seem to have great applicability in effluent treatment and bioremediation in which toxic dyes, azo, phenolic, other aromatic compounds, etc are in soluble phase. Various oxidoreductase enzymes show good efficiency in color removal and bioremediation but lesser stability necessitate to explore these reuse techniques by immobilization (Shukla and Jajpura 2005).

14 Challenges and Future Prospects of Enzyme Applications

In present scenario due to globalization there is cut throat completion amongst the textile industries which necessitates bringing down the cost of production. Enzymes are best alternative to various toxic chemicals being used in textile and apparel industries. Thus they have to compete with these low cost chemicals for their replacement. Besides these their storage and operation stability, resistance towards chemicals and additives used in enzymatic operation have to be addressed seriously. Even hardness, heavy metals, salts, chelating agents present in water also affect the enzymatic operations. The advancements in biotechnology made it possible to cut down the cost of mass production of enzymes as well as gene and protein engineering improved the enzymes stability and resistance towards the harsh conditions but still lot have to be done so that enzymes can be used even in small and medium scale industries.

In certain wet processing steps such as in bioscouring, the larger molecular size of enzymes hinder their penetration inside the interior crystalline regions of the fiber. Hence efficiency of scouring in terms of wax and fat removal has to be compromised as compared with conventionally used scouring with sodium hydroxide. The enzyme penetration using appropriate compatible surfactant can give satisfactory level of scouring with ecofriendly approach.

Amylases, lipases, proteases, catalases and cellulases enzymes have proved themselves in various textile wet processing application efficiently. But the oxidoreductase enzymes such as laccases, oxidases, etc applications in biobleaching, indigo fading is yet to be popularized and made feasible for industrial purpose. The removal of various dyes, azo, phenolic and other toxic compounds from effluent is dire needed in textile and apparel industries. The removal of these compounds from the effluent is tedious and costly process. Thus many industries drain them as it is in the surrounding areas. The laccases, oxidases, etc enzymes can play major role in cleaning up the effluent but availability of these enzymes at low cost in industrial scale is great challenge. The cost and toxicity associated with synthetic mediators limits their applications in enzyme reactions. The ecofriendly bio mediators which govern the enzymatic reactions in plants and other living species can replace the toxic mediators but their extraction methodology is yet to be developed for mass scale production.

In certain cases, bioremediation of toxic hydrophobic aromatic compounds need solvent for enzymatic reaction. Selection of appropriate low cost solvent is again tedious task as inappropriate solvents may affect the enzyme activity severely. Definitely free enzymes have certain limitations in effluent treatment and bioremediation but the immobilized enzyme and whole microbial cells can be explored further for efficient effluent treatment of textile and apparel industries.

Overall enzymes have tremendous scope in detoxifying the supply chain by replacing various toxic chemicals in textile and apparel sector. They not only detox the supply chain by replacing toxic chemicals in various processing steps but also detoxify the harmful substances in effluent treatment operation and waste apparel disposal. Definitely, with advancements in enzyme technology the use of enzymes will increase significantly in cleaning up textile and fashion sector.

References

Agrawal PB (2005) The performance of cutinase and pectinase in cotton scouring. Dissertation, University of Twente, the Netherlands, Wohrmann Print Service, the Netherlands

Ahlawat S, Dhiman SS, Battan B, Mandhan RP, Sharma (2009) J Pectinase production by *Bacillus subtilis* and its potential application in biopreparation of cotton and micropoly fabric. Process Biochem 44(5):521–526

Ajgaonkar DB, Talukdar MK, Wadewkar VR (1982) Sizing materials and methods machines. Textile Trade Press, Ahmedabad, p 5

Akin DE, Rigsby LL, Perkins W (1999) Quality properties of flax fibres retted with enzymes. Text Res J 69(10):747–753

Akin DE, Dodd RB, Perkins W, Henriksson G, Eriksson KEL (2000) Spray enzymatic retting: a new method for processing flax fibres. Text Res J 70(6):486–494

Alinsafi A, Khemis M, Pons MN et al (2005) Electro-coagulation of reactive textile dyes and textile wastewater. Chem Eng Process; Proc Intens 44:461–470

Araujo R, Casal M, Cavaco-Paulo A (2008) Application of enzymes for textile fibres processing. Biocatal Biotransform 6(5):332–349

Bahi A, Jones JT, Carr CM, Ulijn RV, Shao J (2007) Surface characterization of chemically modified wool. Text Res J 77(12):937–945. doi:10.1177/0040517507083520

Banci L, Ciofi-Baffoni S, Tien M (1999) Lignin and Mn peroxidase-catalyzed oxidation of phenolic lignin oligomers. Biochemistry 38:3205–3210

Batra SH (1985) Other long vegetable fibers: abaca, banana, sisal, henequen, flax, ramie, hemp, sunn, and coir. In: lewin M, Pearce EM (ed) Handbook of fiber science and technology. Fiber chemistry, vol 4. Marcel Dekker, New York

Bernath FR, Vieth WR (1974) Immobilized enzyme in food and microbial processes. Plenum Press, New York

Bhavan S, Rao JR, Nair BU (2008) A potential new commercial method for processing leather to reduce environmental impact. Environ Sci Pollut Res 15(4):293–295

Bizuneh A (2012) Textile effluent treatment and decolorization techniques—a review. Chem: Bul J Sci Educ 21(3):434–436

Boyer PD (ed) (1959) Handbook of enzymes, vol I. Acad. Press, New York

Brühlmann F, Kim KS, Zimmerman W, Fiechter A (1994) Pectinolytic enzymes from actinomycetes for the degumming of ramie bast fibers. Appl Environ Microbiol 60(6):2107–2112

Brühlmann F, Leupin M, Erismann KH, Fiechter A (2000) Enzymatic degumming of ramie bast fibers. J Biotechnol 76:43–50

Buchert J, Pere J, Puolakka A, Nousiainen P (2000) Scouring of cotton with pectinases, proteases and lipases. Text Chem Color Am Dyest Rep 32(5):48–52

Campos R, Kandelbauer A, Robra KH, Cavaco-Paulo A, Gübitz GM (2001) Indigo degradation with purified laccases from *Trametes hirsute* and *Sclerotium rolfsii*. J Biotechnol 89:131–139

Cardamone JM (2007) Enzyme-mediated cross linking of wool. Part: Transglutaminase. Text Res J 77(4):214–221

Cavaco-Paula A, Almeida A (1996) Kinetic parameters measured during cellulase processing of cotton. J Text Inst 87:227–233

Cavaco-Paulo A, Gübitz G (2003) Catalysis and processing. In: Cavaco-Paulo A, Gubitz G (eds) Textile processing with enzymes, 1st edn. Woodhead Publishing Limited, Cambridge, pp 86–119

Cavaco-Paulo A, Morgado J, Almeida L, Kilburn D (1998) Indigo backstaining during cellulase washing. Text Res J 68(6):398–401

Ceria A, Rovero G, Sicardi S, Ferrero F (2010) Atmospheric continuous cold plasma treatment: thermal and hydrodynamical diagnostics of a plasma jet pilot unit. Chem Eng Process 49 (1):65–69. doi:10.1016/j.cep.2009.11.008

Chakraborty JN, Jaruhar P (2014) Dyeing of cotton with sulphur dyes using alkaline catalase as reduction catalyst. Indian J Fibre Text Res 39:303–309

Chance B, Sies H, Boveris A (1979) Hydroperoxide metabolism in mammalian organs. Physiol Rev 59(3):527–605

Chesson A (1980) Maceration in relation to the post handling and processing of plant material. J Appl Biotechnol 48:1–45

Chiba S (1997) Molecular mechanism in (X-glucosidase and glucoamylase). Biosci Biotech Bioc/zell 61(8):1233–1239

Claus H (2003) Laccases and their occurrence in prokaryotes. Arch Microbiol 179:145–150

Conn EE, Stumpf PK, Bruening G, Doi RH (1987) Outlines of biochemistry, 5th edn. Wiley, Singapore, pp 115–164

Cortez J, Bonner PLR, Griffin M (2004) Application of transglutaminases in the modification of wool textiles. Enzyme Microb Technol 34:64–72

Couto SR, Toca-Herrera JL (2006) Lacasses in the textile industry. Biotechnol Mol Biol Rev 1 (4):115–120

Crueger W, Crueger A (2000) Biotechnology: a textbook of industrial microbiology, 2nd edn. Panima Publishing Corp., New Delhi, p 189

Damsus T, Kirk O, Pedersen G, Venegas MG (1991) Novo Nordisk A/S, The Procter & Gamble Company, Patent O9105839

Das Gupta PC, Sen K, Sen SK (1976) Degumming of decorticated ramie for textile purposes. Cell Chem Technol 10:285–291

De Souza FR, Gutterres M (2012) Application of enzymes in leather processing: a comparison between chemical and coenzymatic processes. Braz J Chem Eng 29(3):473–481

Dhillon A, Gupta JK, Jauhari BM, Khanna S (2000) A cellulasepoor, thermostable, alkalitolerante xylanase produced by *Bacillus circulans* AB 16 grown on rice straw and its application in biobleaching of eucalyptus pulp. Biores Technol 73:273–277

Drauz K, Waldmann H (1994) Enzyme catalysis in organic synthesis: a comprehensive handbook. VSH, Weinheinm

Eggert C, Temp U, Dean JFD, Eriksson KEL (1996) A fungal metabolite mediates degradation of non-phenolic lignin structures and synthetic lignin by laccase. FEBS Lett 391:144–148

Etters JN (1995) Advances in indigo dyeing: implication for the dyer, apparel manufacturer and environment. Text Chem Color 27(2):17–22

Feitkenhauer H, Fischer D, Fah D (2003) Microbial desizing using starch as model compound: enzyme properties and desizing efficiency. Biotechnol Prog 19:874–879

Freddi G, Mossotti R, Innocenti R (2003) Degumming of silk fabric with several proteases. J Biotechnol 106:101–112

Gornall AG, Bardawill CJ, David MM (1949) Determination of serum proteins by means of the biuret reaction. J Biol Chem 177(2):751–766

Gulrajani ML, Agarwal R, Chand S (2000a) Degumming of silk with fungal protease. Indian J Fibre Text Res 25:138–142

Gulrajani ML, Agarwal R, Grover A, Suri M (2000b) Degumming of silk with lipase and protease. Indian J Fibre Text Res 25:69–74

Hartzell MM, Hsieh YL (1998) Enzymatic scouring to improve cotton fabric wettability. Text Res J 68(4):233–241

Hasan F, Shah AA, Hameed A (2006) Industrial applications of microbial lipases. Enzyme Microb Technol 39(2):235–251

Hasan F, Shah AA, Javed S, Hameed A (2010) Enzymes used in detergents: lipases. Afr J Biotech 9(31):4836–4844

Hedin PA, Jenkis JN, Parrot WL (1992) Evaluation of flavonoids in *Gossypium arboretum* (L.) cottons as potential source of resistance to tobacco budworm. J Chem Ecol 18:105–114

Heine E, Ruers A, Hocker H (2000) Enzymatic degradation of vegetable residues in wool. DWI Rep 123:475–479

Herbots I, Kottwitz B, Reilly PJ et al (2008) Enzymes, non-food application. Ullmann's encyclopedia of industrial chemistry. Wiley-VCH Verlag GmbH & Co. KGaA, Weinheim

Hickman WS (1995) Preparation; cellulosic dyeing. In: Shore J (ed) Society of dyers and colourists. The Alden Press, Oxford

Ibrahim NA, El-Hossamy M, Morsy MS, Eid BM (2004) Development of new eco-friendly options for cotton wet processing. J Appl Polym Sci 93:1825–1836

Industrial Enzymes Market by Type (Amylases, Cellulases, Proteases, Lipases, and Phytases), Application (Food & Beverages, Cleaning Agents, and Animal Feed), Source (Microorganism, Plant, and Animal), and Region—Global Forecast to 2022 (2016) http://www.marketsand markets.com/Market-Reports/industrial-enzymes-market-237327836.html. Accessed 15 Mar 2017

Jajpura L (2014) Decolourisation of textile effluent by laccase—a review. In: Proceeding of International conference—emerging trends in traditional and technical textiles (ICETT), Department of Textile Technology, NIT Jalandhar, 11–12 April 2014

Jajpura L (2015) Sustainable fibre production and textile wet processing for better tomorrow. In: Asian textile conference (ATC-13), 2015 at Deakin University, Wauran Ponds 3216, Geelong, Victoria Australia from 3–6 Nov 2015

Jajpura L, Singh B (2015) Impact of agricultural technologies employed for food and textile fibres production on environment and human health. Environ We Int J Sci Technol 10:101–116

Jajpura L, Khandual A, Pai RS (2004) Effluent treatment in textile industries. Text Mag 4(45): 34–40

Jenkins RO (2003) Enzymes. In: Cavaco-Paulo A, Gubitz G (eds) Textile processing with enzymes, 1st edn. Woodhead Publishing Limited, Cambridge, pp 1–41

Karapinar E, Sariisik MO (2004) Scouring of cotton with cellulases, pectinases and proteases. Fibres Text East Eur 12:79–82

Khandual A, Jajpura L, Pai RS (2004) Sizing processes and its application. Colourage 51(11): 33–40

Kumar A, Mee-Young Y, Purtell C (1997) Optimizing the use of cellulase enzymes in finishing cellulosic fabrics. Text Chem Color 29(4):37–42

Lantto R, Schänberg C, Buchert J (2004) Effects of laccase-mediator combination on wool. Text Res J 74:713–717

Lenting HBM, Warmoeskerken M (2001) Mechanism of interaction between cellulase action and applied shear force, an hypothesis. J Biotechnol 89(2–3):217–226

Li Y, Hardin IR (1998) Enzymatic scouring of cotton-surfactants, agitation and selection of enzymes. Text Chem Color 30:23–29

Li S, Yang X, Yang S, Zhu M, Wang X (2012) Technology prospecting on enzymes: application, marketing and engineering. Comput Struct Biotechnol J 2(3). http://dx.doi.org/10.5936/csbj. 201209017

Liu W, Chao Y, Yang X, Buo H, Qian S (2004) Biodecolorization of azo, anthraquinonic and triphenylmethane dyes by white-rot fungi and a laccase-secreting engineered strain. J Ind Microbiol Biotechnol 31:127–132

Maehly AC, Chance B (1954) The assay of catalase peroxidase. In: Glick D (ed) Methods of biochemical analysis, vol 1. Interscience Publishers Inc., New York, p 357

Mather RR, Wardman RH (2011) The chemistry of textile fibres. RSC Publishing, Cambridge

McNeil M, Darvill AG, Fry SC, Albertsheim P (1984) Structure and function of the primary cell walls of plants. Ann Rev Biochem 53:625–663

Messing RA (1975) Immobilized enzyme for industrial research. Acad Press, New York

Metzler DE (2001) Biochemistry the chemical reactions of living cells, vol I. Harcourt Academic Press, San Diego, pp 459–466

Michael S, William PG (1997) The mechanism of hydrogen peroxide bleaching. Text Chem Color 29:11

Michaelis L, Menten ML (1913) Die Kinetik der Invertinwirkung. Biochemische Zeitschrift 49:333–369

Mohanty AK, Manjusri M, Drzal LT (2005) Natural fibres, biopolymers and biocomposites. CRC Press, Taylor & Francis Group, Boca Raton

Mojsov K (2011) Application of enzymes in the textile industry: a review. In: 2nd International Congress "Engineering, Ecology and Materials in the Processing Industry" Jahorina, 9–11 March 2011, pp 231–239

Mojsov K (2012) Enzyme scouring of cotton fabrics: a review. Int J Mark Technol 2(9):256–275

Morgado J, Cavaco-Paulo A, Rousselle M (2000) Enzymatic treatment of lyocell-clarification of depilling mechanisms. Text Res J 70(8):696–699

Morton WE, Hearle JWS (1993) Physical properties of textile fibres, 3rd edn. Textile Institute, Manchester

Mousa AHN (1976) Optimisation of rope-range bleaching of cellulostic fabrics. Text Res J 46:493–496

Nielsen PH, Kuilderd H, Zhou W, Lu X (2009) Enzyme biotechnology for sustainable textiles. In: Blackburn RS (ed) sustainable textiles. Woodhead Publishing, Cambridge, pp 113–138

Nomenclature Committee of The International Union of Biochemisty and Molecular Biology (NC-IUBMB) (1992) Enzyme nomenclature. Academic Press, San Diego

Othmer K (1980) Encyclopedia of chemical technology, vol 9. p 138

Palmer T (1981) Understanding enzymes. Ellis Horwood Ltd., New York, p 17

Pedersen AH, Schneider PNN (1998) US Pat. 5795855 A. US-Patent, 1998

Pereira L, Bastos C, Tzanov T, Cavaco-Paulo A, Gübitz GM (2005) Environmentally friendly bleaching of cotton using laccases. Environ Chem Lett 3(2):66–69

Peters RH (1967) Textile chemistry, vol II. Elsevier Publishing Company, London, p 150

Rajamani S, Chen Z, Zhang S, Su C (2009) Recent developments in cleaner production and environment protection in world leather sector. In: 30th IULTCS Congress 2009, Beijing, China, p 5

Ramachandran P, Sundharam R, Palaniyappan J, Munusamy AP (2013) Potential process implicated in bioremediation of textile effluents: a review. Pelagia Res Libr, Adv Appl Sci Res 4(1):131–145

Rangi A, Jajpura L (2015) The biopolymer sericin: extraction and applications. J Text Sci Eng 5 (1):1–5

Rau M, Heidemann C, Pascoalin AM et al (2008) Application of cellulases from *Acrophialophora nainiana* and *Penicillium echinulatum* in textile processing of cellulosic fibers. Biocatal Biotransform 26(5):383–390

Richardson S, Gorton L (2002) Characterisation of the substituent distribution in starch and cellulose derivatives. Analytica Chimica Acta 497:27–65

Rinsey JVA, Karpagam CS (2012) Degumming of silk using protease enzyme from bacillus species. Int J Sci Nat 3(1):51–59

Rodwell VW, Kennelly PJ (1999) Enzymes kinetics. In: Harper's biochemistry, A lange medical book, 25th edn. Appleton & lange, Stamford, pp 86–102

Rott U, Minke R (1999) Overview of wastewater treatment and recycling in the textile processing industry. Water Sci Technol 40:37–144

Sakai T, Sakamoto T, Hallaert J, Vandamme EJ (1993) Pectin, pectinase and protopectinase: production, proterties and applications. Adv Appl Microbiol 39:213–294

Sangwatanaroj U, Choonukulpong K (2003) Cotton scouring with pectinase and lipase/protease/cellulase. AATCC Rev 3:17–20

Sarkar AK, Etters JN (1999) International Conference and Exhibition, AATCC, 12–15 Oct 1999, p 274

Schimper CB, Constanta I, Bechtold T (2009) Effect of alkali pre-treatment on hydrolysis of regenerated cellulose fibers by cellulases (part 1: viscose). Cellulose 16(6):1057–1068. doi:10. 1007/s10570-009-9345-6

Setti L, Giuliani S, Spinozzi G, Pifferi PG (1999) Laccase catalyzedoxidative coupling of 3-methyl 2-benzothiazolinone hydrazone and methoxyphenols. Enzyme Microb Technol 25:285–289

Shao J, Hawkyard CJ, Carr CM (1997) Investigation into the effect of UV/ozone treatments on the dyeability and printability of wool. J Soc Dyers Colour 113(4):126–130. doi:10.1111/j.1478-4408.1997.tb01884.x

Shenai VA (1991) Technology of bleaching and mercerisation, 2nd edn. Mumbai, Sevak Pub., p 37

Shin HS, Guebitz G, Cavaco-Paulo A (2001) In situ enzymatically prepared polymers for wool coloration. Macromol Mater Eng 286:691–694

Shukla SR, Jajpura L (2004) Estimating amylase activity for desizing by DNSA. Text Asia, 15–20

Shukla SR, Jajpura L (2005) Immobilisation of amylase by various techniques. Indian J Fibres Text Res 3(29):75–81

Shukla SR, Maheshwari KC (2002) Use of standing bath technique in peroxide bleaching of cotton. Color Technol 118(2):75–78

Shukla SR, Jajpura L, Damle AJ (2003) Enzyme: the biocatalyst for textile processes. Colourage, Special issue on TEXTINDIA FAIR Club Melange, 7–9 Nov 2003, pp 41–47

Silva CM, Carneiro F, O'Neill A, et al (2005) Cutinase—a new tool for biomodification of synthetic fibers. J Polym Sci Part A: Polym Chem 43:2448–2450

Silva C, Araújo R, Casal M, Gubitz GM, Cavaco-Paulo A (2007) Influence of mechanical agitation on cutinases and protease activity towards polyamide substrates. Enzyme Microb Technol 40:1678–1685

Singh R, Kumar M, Mittal A, Mehta PK (2016) Microbial enzymes: industrial progress in 21st century. 3 Biotech 6:174–175. doi:10.1007/s13205-016-0485-8

Smith JE (1996) Biotechnolgy, 3rd edn. Cambridge University Press, Cambridge, pp 68–83

Snyder LG (1997) Improving the quality of 100% cotton knit fabrics by defuzzing with singeing and cellulase enzymes. Text Chem Color 29(6):27–31

Sumner JB (1926) The isolation and crystallization of the enzyme urease: preliminary paper. J Biol Chem 69:435–441

Suskling CJ (1984) Enzyme chemistry. Chapman and Hall, London

Svendsen (2000) Lipase protein engineering. Biochim Biophys Acta 1543:223–238

Taylor RF (1991) Protein immobilization: fundamentals and application. Marcel Dekker, Inc., New York

Torres E, Bustos-Jaimes I, Borgne SL (2003) Potential use of oxidative enzymes for the detoxification of organic pollutants. Appl Catal B: Environ 46:1–15

Traore MK, Buschle-Dilleer G (2000) Environmentally friendly scouring processes. Text Chem Color Am Dyest Rep 32(12):40

Trotman ER (1968) Textile scouring & bleaching. Griffin Publishers, London, p 33

Trotman ER (1984) Dyeing and chemical technology of textile fibers, 6th edn. Charles Griffin and Company Ltd., High Wycombe

Tyndall RM (1992) Improving the softness and surface appearance of cotton fabrics and garments by treatment with cellulase enzymes. Text Chem Color 24(6):23–26

Tzanov T, Basto C, Gübitz GM, Cavaco-Paulo A (2003a) Laccases to improve the whiteness in a conventional bleaching of cotton. Macromol Mater Eng 288:807–810

Tzanov T, Silva CJ, Zille A, Oliveira J, Cavaco-Paulo A (2003b) Effect of some process parameters in enzymatic dyeing of wool. Appl Biochem Biotech 111:1–14

Van Rensburg NJJ, Barkhuysen FA (1983) Continuous shrink-resist treatment of wool tops using chlorine gas in a conventional suction-drum backwash. SAWTRI Tech Rep 539:22

Van Sumere C, Sharma H (1991) Analysis of fine flax fiber produced by enzymatic retting. Aspect Appl Biol 28:15–20

Varanasi A, Obendorf SK, Pedersen LS, Mejldal R (1997) Lipid distribution on textiles in relation to washing with lipases. J Surfactants Deterg 4:135–146

Walker GM, Weatherly LR (1997) Adsorption of acid dyes onto granular activated carbon in fixed beds. J Water Res 31:2093–2101

Weil JH (1996) General biochemistry, 6th edn. New Age International Limited, New Delhi

West ES, Todd WR (1957) Textbook of biochemistry, 2nd edn. The Macmillan Company, New York, pp 411–460

Wu Z, Joo H, Ahn IS, Haam S, Kim JH, Lee K (2004) Organic dye adsorption on mesoporous hybrid gels. Chem Eng J 102:277–282

Xia J, Chen X, Nnanna IA (1996) Activity and stability of Penicillium cyclopium lipase in surfactant and detergent solutions. J Am Oil Chem Soc 73:115–120

Yoon MY, McDonald H, Chu K, Garratt C (2000) Protease, a new tool for denim washing. Text Chem Color Am Dyest Rep 32(5):25–29

Yoon MY, Kellis J, Poulose AJ (2002) Enzymatic modification of polyester. AATCC Rev 2 (6):33–36

Zahn H (1993) Silk. Ullmann's encyclopedia of industrial chemistry, vol A24. VCH Publisher Inc., pp 95–106

Printed in the United States
By Bookmasters